Beyond the Veil: Unraveling Reality's Code

by Aqueel Premjee

A Quest for the Ultimate Nature of Reality

This is a work of nonfiction. While speculative in nature, all interpretations are grounded in scientific theory, philosophical reflection, and the author's personal insights.

Cover design by Aqueel Premjee (in collaboration with AI tools)

Dedication

For those who question the frame they were born into— and dare to look beyond the veil.

Acknowledgements

This book was shaped by countless minds—known and unknown—who lit the way with questions, theories, and truths wrapped in poetry. I thank the giants whose shoulders I stand on, the friends who challenged my thinking, and the quiet moments that whispered clarity. To the readers walking beside me now: may this journey echo yours.

Special thanks to my family, mentors, and all seekers of the unseen

Preface

The Personal Spark

My quest for the nature of reality has been a winding one, kindled by a lifelong obsession wanting to determine my true purpose. My journey into this realm of thought began at the age of fourteen, amidst the rigid doctrines of my religious upbringing. I was told that my faith was the sole path to truth, that all others were misguided. But a nagging question persisted: what if I had been born into a different family, a different culture—would I not be singing a different song, believing in a different truth? This dissonance ignited a quest for understanding, a search for a reality that transcended the confines of dogma. This quest led me to the study of physics, a discipline that seeks to unravel the fundamental laws of the universe. Through undergraduate and graduate school, I immersed myself in the intricacies of quantum mechanics, drawn to the groundbreaking work of figures like Bohr, Dirac, and Fermi. I was captivated by Einstein's revolutionary theories of relativity and Feynman's elegant explanations of quantum phenomena. These were not just great scientists, but also great thinkers, unafraid to challenge conventional wisdom. Max Tegmark's bold vision of a mathematical universe further fueled my curiosity, prompting me to question the very fabric of reality. As Newton said, 'If I have seen further, it is by standing on the shoulders of giants.' Scientists, prophets, philosophers—their insights illuminate reality's vast tapestry, guiding my quest to unravel its code.

When I first encountered the question if it is possible that we are living in a simulation, it struck me as a far-fetched idea—perhaps a 1% chance, a reluctant nod to the irrefutability of the unknown. Yet, as science continued to unveil the mysteries of the cosmos—refinements to cosmic inflation theory, precision mapping of the cosmic microwave background, the rise of deep learning and artificial intelligence, the confirmation of dark energy and the Higgs boson, the discovery of exoplanets, and the detection of gravitational waves—each breakthrough felt like a veil lifting, revealing deeper layers of the code beneath reality. Long before I knew his stance, I'd drift into daydreams of a conversation with Elon Musk—probing him on simulation theory, wondering if he'd deem it possible or nudge me back to realism's firm ground. Later, when videos of him declaring we could likely be in a simulation flickered across my screen, a flame flared—his words not only echoed my musings but lifted them higher, deepening my admiration for a mind unafraid to chase the cosmos's edges. Donning a virtual headset and experiencing the advent of virtual reality with its immersive simulations further fueled my curiosity. If such convincing worlds could already be created—spaces to walk through, objects to touch, gravity to feel—I found myself wondering how much more real they would seem in just a decade or two. Would the senses, so easily fooled even now, eventually yield completely to the illusion?

I immersed myself in the writings of great thinkers—those bold enough to question the foundations of reality itself. I didn't always understand every equation or concept, but I was hooked by the sheer audacity of their ideas. How could time stretch or space bend? How could reality be anything other than what it seemed? That curiosity became a compass, nudging me ever deeper into the strange and beautiful world of physics. But it was Richard Feynman who first lit the flame, declaring, "Physics is the ultimate intellectual adventure." That phrase captured the thrill of the chase—the idea that unraveling the universe's mysteries was a grand, exhilarating quest. I devoured his *Lectures on Physics* volumes and *Surely You're Joking, Mr. Feynman!*, utterly impressed by his clarity, humor, and boundless curiosity. His ability to distill profound truths into elegant simplicity made him a beacon in my intellectual journey. Yet beyond the precision of his equations, Feynman revealed a deeper poetry hidden in the mechanics of nature.

One particular reflection sealed his genius for me:

"You see, the trees are made of air, primarily. When they're burned, they go back to air, and in the flaming heat is released the energy of the sun which was bound in to convert the air into tree, and in the flame, you see the energy being released, and it's the same energy that was used to make the tree in the first place. And the ashes that remain—that's the part that didn't come from the air, but from the solid earth."

I lingered over those words, captivated by how he wove the tree's life into a poetic cycle of air, sun, and earth, unveiling the universe's hidden elegance—its conserved energies and balanced harmonies—with a clarity that felt almost magical.

Then came a moment I'll never forget—a privilege and an honor that stands out as one of the most unforgettable experiences of my life. In 1987, just a year before his passing, I had the extraordinary chance to hear Feynman speak at UC Irvine. It was a lecture on quantum electrodynamics, delivered shortly after he'd published *QED: The Strange Theory of Light and Matter*. Sitting there, I watched in awe as he unraveled the quirks of light and matter with that signature blend of wit and wisdom, making the incomprehensible feel within reach. Afterward, I joined the crowd eager to approach him, and when my turn came, I shook his hand. That brief clasp—his warm grip, the glint in his eye—felt like a bridge to the cosmos itself. To meet the man whose words had ignited my passion for physics, whose mind had peeled back the layers of reality, was a humbling gift. It was a fleeting encounter, yet it left an indelible mark, fueling my drive to explore the unknown.

That encounter with Feynman crystallized something I had long felt but struggled to articulate—that behind the precision of science lies a current of wonder, and that seeking truth is as much a personal calling as it is an intellectual pursuit. What began as a search to reconcile inherited belief with observable reality grew into a lifelong exploration of deeper patterns—both physical and metaphysical. Today, I find myself standing at the threshold of a radical question: *What if reality itself is a rendered illusion?* Not a fantasy, but a coded construct—one where consciousness is not merely a byproduct, but perhaps the reason it all exists.

And so, I invite you to join me. Let us begin where wonder meets inquiry, and follow the trail where it leads—beyond the veil.

Prologue

Between Code and Cosmos

What is the nature of reality?

It's a question that has echoed through temples and laboratories, through the quiet hum of neurons and the radiant arcs of galaxies. Is the world outside of us truly "out there," or are we inhabiting a curated illusion—crafted either by divine hand, natural law, or intelligent design?

As a physicist, I've long been drawn to the elegance of equations and the stubborn certainty of measurements. But reality, it seems, does not sit still. The deeper we look, the more it slips between our fingers.

Quantum mechanics tells us that particles exist in states of probability until observed. Einstein, puzzled by its implications, famously quipped, "I like to think the moon is there even if I am not looking at it." Yet Bell's Theorem and the experiments that followed have shown that hidden variables may not restore realism. Reality is not local, not definite—at least not in the way we intuitively expect.

Simultaneously, advances in information theory, artificial intelligence, and digital physics—the idea that the universe may be composed of discrete informational units—have led to a provocative possibility: that the universe itself may be a simulation—a rendered environment computed at extraordinary scales by a system far more complex than our own minds. What was once science fiction is now taken seriously by leading thinkers.

But to call the universe a simulation is not necessarily to diminish it. A simulation can be beautiful. A simulation can contain real consciousness. The question isn't just whether we are simulated—but whether we can understand the code.

In this book, I invite you to join a journey that straddles physics and philosophy, science and story. We'll explore the logic of simulated universes, the enigma of consciousness, and the surprising intersections of ancient texts with modern theories.

Along the way, I'll share reflections and thought experiments, threading personal moments with the puzzle of existence. This is not a book of answers, but a map of questions. Some may unsettle. Others may ignite a sense of wonder you thought long lost.

We begin not with a conclusion—but with a veil. And beyond it, perhaps, something astonishing awaits.

Chapter 1: The Fabric of Reality - Evidence from Physics

Introduction

The universe pulses with a strange elegance—stars spiral through vast galactic arms, atoms spin in unseen dances, and time bends under gravity's subtle pull. From a falling apple's arc to the glow of distant galaxies, physics reveals a cosmos both familiar and wondrous, governed by laws that dazzle with their order yet hint at deeper riddles. Why does reality seem so finely tuned, its smallest pieces eluding our grasp, its balance cradling life on a cosmic knife's edge? This chapter explores that wonder, tracing science's path from a clockwork past to the frontiers of quantum and cosmic mysteries. Are these mere quirks of nature, or signs of a design too vast to fathom?

A Clockwork Cosmos

Long before quantum uncertainty rattled the foundations of science, a vision of perfect predictability held sway. Pierre-Simon Laplace imagined a vast intellect—one that, knowing the position and momentum of every atom in the universe, could calculate the entire past and future with flawless precision. Nothing hidden. Nothing random. A cosmos of gears and trajectories.

This was the clockwork model of the universe. By the late 19th century, some physicists even believed that all fundamental laws had been discovered—that science was entering an age of refinement, not revolution. The great task ahead, they thought, was to fine-tune the equations, measure constants more precisely, and apply known laws to the messy complexity of nature.. A majestic unfolding of cause and effect. Newton's laws gave us the tools. His seminal work, *Mathematical Principles of Natural Philosophy*, laid the foundation for classical mechanics, introducing gravity, force, motion, and inertia—concepts expressed through elegant equations that charted the dance of planets and pendulums. With this framework, the future was no longer a mystery but a calculation waiting to be made, a cosmos unfolding with majestic predictability. But it left little room for choice, creativity, or consciousness. A mind, in such a world, was simply another gear.

And then, the equations faltered.

Quantum mechanics arrived not with a whisper, but a rupture. It didn't tinker with the machine—it revealed that the machine had never existed. That nature, at its root, is governed not by determinism, but probability. That knowledge has limits. That reality dances with the observer.

Laplace's demon, it turns out, could not see everything. And the cosmos may not run on rails, but on riddles.

Direction in a Directionless Universe

Entropy, too, plays its part in this ordered reality. The second law of thermodynamics declares that in any isolated system, entropy — the measure of disorder — will increase over time.

It governs the cooling of stars, the fading of memory, the inevitable disintegration of all things built.

Entropy is not a force, like gravity. It does not push, pull, or radiate. It is a statistical trajectory — a whisper of direction in an otherwise timeless equation.

Entropy flows one way—and with it, time. Yet behind this simple law, might something deeper stir?

The Digital Universe

At the smallest scales, the cosmos shifts from smooth continuity to a stranger realm of finite steps. At the Planck scale, spacetime may fragment into a lattice of discrete points, not an endless expanse. Time, space, energy: these quanta suggest a universe of coded precision, like a vast screen rendering existence pixel by pixel. Such granularity implies boundaries, as if the cosmos spares its details, shaping only what's observed. This digital structure hints at a deeper truth—perhaps a system woven with intent, its limits a shadow of a purposeful design.

The idea of a digital universe suggests that everything—particles, forces, space, even time—emerges from underlying information, processed in accordance with rules we call physical laws. Reality, in this view, is not continuous but discrete. Like pixels forming an image or code compiling into behavior, the cosmos might be computed.

John Archibald Wheeler's phrase 'it from bit' captures this succinctly: every physical 'thing' arises from an immaterial binary choice. A bit—yes or no, on or off—underlies it all.

Early pioneers like Konrad Zuse, Edward Fredkin, and John Archibald Wheeler speculated that the universe itself might be fundamentally computational—an idea now known as digital physics."

It's a shift not just in physics, but in metaphor. We move from mechanism to message. From machine to language.

Could space itself be like a grid? Could the speed of light represent a frame rate? Could the Planck length be a pixel size? In this model, black holes become data vaults, entropy becomes a measure of information loss, and quantum entanglement begins to resemble data sync across a distributed system.

In a simulated universe, rendering only occurs when observed. Complexity appears only as needed. This model would solve the riddle of why vast cosmic structures exist yet remain dark and unseen, and why the laws of physics appear finely tuned for life.

But even if the universe is computed, it doesn't make it fake. Simulated does not mean unreal. A simulation can host real minds, real feelings, real meaning. The question is not whether it's pixels or particles—it's whether we can decode the pattern.

And if the universe is code, then somewhere, there is a coder—or at least, a compiler. Whether natural or intelligent, personal or abstract, the existence of such a system raises profound implications.

Not just about physics.

But about purpose.

Gravity's Elegant Illusion

Among nature's forces, gravity holds a peculiar place. It sculpts galaxies, binds planets in orbit, and gives weight to our every step. Yet of the four fundamental interactions, it is the weakest by far—trillions of times weaker than electromagnetism, and vastly overshadowed by the strong and weak nuclear forces.

So why does gravity dominate our experience?

Einstein's general relativity answered this not by describing gravity as a traditional force, but by reimagining it as a feature of geometry. Massive objects curve spacetime, and everything else—planets, light, you and me—follows those curves. Apples don't fall because of a pull. They fall because the Earth warps the fabric of spacetime.

This geometric view of gravity is breathtaking in elegance. It predicted black holes, gravitational lensing, and the expansion of the universe. It remains one of the most precise theories ever tested. And yet, it refuses to merge with the quantum world.

Attempts to quantize gravity have led to string theory, loop quantum gravity, and holographic models—each offering glimpses, none complete. In some theories, gravity isn't even fundamental but emergent. A byproduct. A residue.

Dutch physicist Erik Verlinde proposed that gravity may arise from the entropic tendency of systems to increase in disorder. In this view, gravity is not a force or a field, but the result of statistical behavior, like temperature or pressure. It's not woven into spacetime—it emerges from the information within it.

If true, gravity becomes the greatest illusion of all: not a fundamental law, but a side effect of deeper rules. A whisper of the code beneath the world.

What we experience as weight, then, might not be a push or pull, but a hint—a subtle reminder—that reality is not what it seems.

Individuation Through Spacetime Geometry

What if space and time are not merely containers for reality, but engines of individuation—mechanisms that carve distinct trajectories for particles, objects, and observers alike?

In relativity, every particle traces a unique worldline through spacetime—a personal path defined by position and time. These worldlines define what a particle experiences, what it interacts with, and how its reality unfolds. Due to relativistic effects, observers in different frames of reference—whether moving at different velocities or within different gravitational potentials—experience time differently. Simultaneity is not absolute; each worldline carries its own unique version of 'now.' There is no universal clock, no absolute present—only timelines, intersecting.

In quantum mechanics, entangled particles challenge these separations, maintaining correlations across distance. But even then, each particle's state evolves along its own timeline. Locality remains meaningful, even when connection seems to defy it.

If spacetime is quantized at the Planck scale, it imposes fundamental limits on how much information can be stored or transferred between regions. These limits may be what define the boundaries between systems—between one timeline and another, one observer and the next.

Individuation, then, need not be mystical. It may be a natural outcome of how spacetime encodes experience. Physics offers no definitive answer yet—but what we call a "self" may simply be a localized trajectory through an information-bound continuum.

Each of us, a thread in the cosmic fabric—separate only by the path we take.

Black Holes: Edges of the Unknown

At the cosmos's extremes, black holes emerge as enigmas, regions where light fades and time slows to a halt. Consuming matter and information, they may define limits where the universe's laws unravel, with mathematical models breaking down and infinities complicating our understanding.

Their event horizons, storing data like a cosmic archive, hint at a surface concealing deeper truths. These boundaries might capture the essence of what vanishes or connect to realms beyond our grasp. Their mystery suggests they mark the cosmos's frontiers, where familiar patterns yield to unanswerable puzzles. This silence evokes a system shaped with purpose, its edges perhaps tracing a framework we cannot yet fathom.

Deletion Nodes - The Black Hole Information Loss Paradox

Yet a deeper disquiet stirs within these cosmic sentinels. According to quantum theory, information must never be destroyed—the universe, at its core, conserves every particle's history, every wave's whisper. Yet black holes, according to Hawking's radiation theory, seem to erase it utterly. As matter falls past the event horizon, its details dissolve, and when the black hole evaporates, the information appears gone—forever lost to the void. This clash—the Black Hole Information Loss Paradox—sets two pillars of physics at odds. In a natural universe, it is a scandal; in a simulated one, it may be a feature. Black holes could serve as deletion nodes within the code, points where data no longer needed is permanently scrubbed from the system's memory. If the simulation must manage finite resources while rendering an endless stage, then black holes may not just be mysteries—they may be maintenance, silent routines that prune the database of existence itself.

The Illusion of Solidity

What if everything you touch is a lie?

We believe we're surrounded by substance—tables, walls, bodies, earth. But physics tells a different story. It tells us that matter is mostly nothing. That the stuff of the world, when magnified to its core, is emptiness held in place by invisible forces.

Atoms are not solid spheres but tiny nuclei orbited by electrons across vast chasms of space. If an atom were the size of a stadium, its nucleus would be a grain of rice at the center, the electrons swirling at the edges, and all the space between, nothing. Solidity, it turns out, is a perceptual illusion—a trick played by electromagnetic repulsion between electrons. Even within the atomic nucleus, quarks—the building blocks of protons and neutrons—are not static occupants, but dynamic pulses, occupying space for only a tiny fraction of the nucleon's volume. You never truly touch anything; you only feel the field pushing back.

Yet this illusion is incredibly convincing. We trust it with our lives. We build civilizations upon it. Still, it reveals a deeper truth: that what feels most real may be fundamentally hollow. That solidity is not the bedrock of reality, but a mirage cast by unseen forces.

Quantum Uncertainties: Reality's Hazy Edge

Dive smaller still, and the cosmos blurs. Electrons drift as clouds of chance, fixed only when observed. Heisenberg's Uncertainty Principle decrees a trade-off: know a particle's place, and its speed slips; track its speed, and its place fades. This isn't error but law—nature's refusal to be pinned. Quarks shimmer in fractal haze, more void within void.

Schrödinger's Cat probes this strangeness: a feline in a box, tied to a radioactive atom with a 50% chance to decay, poison poised to strike. Quantum logic says the atom is both decayed and not, the cat both alive and dead, until a glance decides. Some, like the Copenhagen view, say observation picks one fate; others, like Everett's Many-Worlds, claim both endure, splitting reality's thread. In this interpretation, every possible outcome of a quantum event actually occurs—each in a different branch of the universe. When the atom decays, a version of the cat dies; when it doesn't, another version lives. Rather than collapsing probabilities into one, reality expands into many, each representing a complete and valid world.

Entanglement adds a twist—particles linked across gulfs, one's shift nudging the other instantly, as if space were no barrier. These quirks defy sense. Are they nature's caprice, or signs of a system rendering choices on demand, sparing effort until we look? Quantum mechanics did not just revise our understanding of the small—it reframed the very structure of knowledge. Where classical physics offered prediction, quantum theory delivered probability. Where we once expected certainty, we found only shadow.

The quantum world is one of superpositions, where particles are not here or there, but both—until observed. The famous double-slit experiment revealed a fundamental mystery: particles behave like waves when unobserved, yet as particles when watched. Observation collapses possibilities into one. This isn't just a measurement problem—it's a philosophical rupture. What is a particle before it's seen? Is reality in flux until mind calls it forth? Or is it the measuring apparatus itself—our interaction with the world—that defines the outcome?

The Heisenberg Uncertainty Principle further deepens the riddle: the more precisely we know a particle's position, the less precisely we know its momentum. Nature draws a curtain—not from lack of tools, but as a fundamental rule.

This haze of probabilities is not ignorance—it's built in. Quantum fields buzz with potential, collapsing into form only when forced to choose. And even then, entanglement ensures that choices made here affect states elsewhere.

We are no longer observers standing outside a clockwork universe. We are participants inside a shimmering lattice of uncertainty. Reality, it seems, is not a fixed stage but a shifting superposition—resolved only in context.

And so, physics no longer promises the certainty Laplace once dreamed. Instead, it offers something stranger, and perhaps more honest: a universe where what we know is tied to how we ask, and where every answer carries the ghost of another possibility.

The Constants of Creation

Zoom out, and the cosmos reveals a new kind of mystery—not in chaos, but in precision. Beneath the grandeur of galaxies and the dance of atoms lie values so specific, so delicately balanced, that a fraction of change would rewrite existence. These are the constants: the speed of light, Planck's constant, the mass and charge of the electron. They are not derived from anything deeper—at least, not yet. We measure them. We plug them in. They are the given ingredients in nature's recipe.

Physics recognizes at least 32 fundamental constants that define the behavior of reality, as identified by Tegmark. Many are embedded in the Standard Model—our best description of particles and forces. Among these, the fine-structure constant (alpha) governs the strength of electromagnetic interactions, while the proton-to-electron mass ratio (mu) shapes atomic structure. Neither can be explained from first principles. Their values simply are.

Why these numbers? Why this structure? If gravity were just a touch stronger, stars might burn too fast, collapsing before planets form. A weaker nuclear force, and atoms couldn't bind. A shift in alpha, and chemistry would fall apart. Life, complexity—even stable matter—hinges on the razor's edge.

We hunt these constants for their mathematical elegance, but perhaps they are more than numbers—they are evidence. Not of accident, but intent. A cosmos whose rules permit life, observation, and meaning may not simply exist, but be authored."

Some call it coincidence. Others lean on the anthropic principle: we observe a universe compatible with life because only such a universe permits observers. But that feels less like an answer and more like a shrug.

The tuning is eerie. Almost… deliberate.

If the universe is a program, these constants are its settings—hard-coded before the first frame. They shape the grid, define the rules, and hint that something—not necessarily someone—selected this version of reality.

Whatever their origin, these numbers do not explain themselves. They remain silent, yet central. Max Tegmark has gone further, proposing that the universe may not just contain math—it may *be* math. In his Mathematical Universe Hypothesis, the fabric of reality isn't just described by equations, it *is* equations—mathematical structures that exist independently and manifest as physical reality.

They remain silent, yet central. And in that silence, they suggest that beneath the seeming chaos lies not randomness—but restraint.

Fine-Structure Drift: A Subtle Whisper in the Code

Yet even among these supposedly immutable constants, a whisper of instability stirs. Observations of distant quasars suggest that alpha—the fine-structure constant—may have shifted by a fraction over billions of years. A tuning too faint for ordinary perception, but profound if true. In a natural cosmos, such drift unsettles the notion of fixed laws; in a simulated one, it hints at live code—parameters refined during runtime, updates rippling across epochs. If the constants themselves can flex, even slightly, then the cosmos we inhabit may be less a finished artifact than a living program, continuously calibrated to sustain its story.

The Earth-Moon Alignment: A Celestial Signature?

The Moon drifts in Earth's sky, a silent partner in a cosmic dance that feels too precise for chance. Its diameter is dwarfed by the Sun, a blazing giant 400 times wider. Yet, from our vantage, their apparent sizes converge with uncanny symmetry—both spanning roughly half a degree—because the Sun lies 400 times farther away. This exquisite balance crafts total solar eclipses, where the Moon cloaks the Sun's disk, unveiling the corona's ghostly halo, a spectacle unmatched elsewhere in our solar system. Nudge the Moon's orbit even slightly, and this harmony falters; eclipses would blur into partial shadows or vanish. Such a tailored alignment feels like a cosmic flourish, a detail etched to stir conscious minds. In a simulated reality, might this be a deliberate cue—a coded whisper prompting us to question the fabric of our world?

This precision shapes more than spectacle. The Moon's gravitational pull anchors Earth's axial tilt at about 23.4 degrees, ensuring seasons unfold steadily, a rhythm life depends on. Without the Moon, Earth's tilt would vary chaotically over time — anywhere between 0° and 85° over millions of years, due to gravitational nudges from the Sun and other planets (mostly Jupiter). This would cause:

- Wild seasonal extremes
- Periods of no seasons at all
- Extreme shifts in climate patterns
- Huge challenges for life to adapt and evolve consistently

Because of the Moon's significant mass and its gravitational interaction with Earth, it dampens these variations, keeping the tilt relatively stable — fluctuating just a little between 22.1° and 24.5° on a 41,000-year cycle.

These tides stir oceans, blending nutrients that feed marine ecosystems, life's ancient cradle. These effects, so delicately poised, suggest more than luck.

The Universe's Calibration: A Tapestry of Intent?

The universe unfolds with an elegance that borders on the uncanny, its every layer humming with a balance too intricate for blind chance. Born 13.8 billion years ago in a flash of unfathomable heat, it expanded not too swiftly to scatter into nothingness, nor too slowly to collapse under its own weight. This equilibrium, set in the Big Bang's first moments, allowed galaxies to coalesce, stars to ignite, and planets to form—each a stepping-stone for life's fragile possibility.

Consider the strength of gravity: if it were a fraction stronger, stars would burn out too quickly for biology to take root; a touch weaker, and matter would drift apart, starless. The cosmic microwave background, a snapshot of the universe's infancy, reveals fluctuations tuned to one part in 100,000—Precise enough to seed structure, yet not so wild as to tear it apart, these calibrations, repeated across countless parameters, feel like a symphony scored for meaning, raising a profound question: Is this a cosmic fluke, or a framework crafted to host conscious minds?

This orchestration deepens in the realm of stellar alchemy. Stars forge elements through fusion, but the process hinges on a delicate resonance in helium's nuclear dance, enabling carbon to form in abundance—a cornerstone for organic molecules. A slight shift in this resonance, and carbon's yield would plummet, unraveling the chemistry of life.

Beyond physics, the universe's initial conditions defy entropy's chaos. Why did it begin so ordered, with energy so evenly spread, allowing complexity to emerge over eons? Theoretical models suggest alternative universes could choke on disorder or stall in uniformity, yet ours threads a needle, fostering structures that think and dream.

Recent studies, like those probing the universe's curvature (2024 Planck data), find it flat to within 0.2%, a geometry ideal for stability—another dial set just so. In a simulated cosmos, these might be no accidents but choices: parameters coded to render a stage where awareness can flourish. What if the Coder's art lies not in chance but in precision, each law a brushstroke painting a reality where minds probe its seams?

Perhaps the universe itself is a question, its fine-tuning a prompt for us to seek the hand behind the code.

The Mathematical Universe

Why does mathematics so precisely mirror the fabric of reality? In 1623, Galileo Galilei proclaimed the universe a grand book inscribed in the language of circles and triangles, its truths laid bare through geometric forms. Centuries later, in 1960, Eugene Wigner marveled at mathematics' "unreasonable effectiveness," its uncanny ability to not only describe but unveil the deepest truths of nature. Consider the Fibonacci sequence—0, 1, 1, 2, 3, 5, 8, 13, 21, 34, 55, and beyond—where each number is the sum of the two preceding ones. This sequence converges toward the golden ratio ($\varphi \approx 1.6180339887$), a mathematical constant defined as the number where the ratio of two quantities equals the ratio of their sum to the larger quantity $(a+b)/a = a/b \approx 1.618$. As the Fibonacci numbers grow, the ratio of consecutive terms—such as 55/34 or 34/21—approaches φ with increasing precision, embodying a proportion that manifests in nature's designs. Sunflowers display this in their 34 and 55 spiral seed patterns, pinecones in their 5, 8, or 13 scale arrangements, and nautilus shells in their logarithmic spirals. Even DNA's structure reflects this harmony, with its double helix measuring approximately 3.4 nanometers wide and completing a half-turn every 2.1 nanometers, yielding a ratio evocative of φ. Beyond life, hurricanes swirl and galaxies spiral in patterns that echo this mathematical rhythm, their forms curling toward the same proportion. Max Tegmark boldly posits that reality itself is mathematics, its fundamental constants—such as the fine-structure constant or the proton-to-electron mass ratio—serving as the rules of a cosmic game. Why this profound alignment? Does mathematics merely describe the world, capturing its patterns like a map, or does it define it, with existence unfolding as a calculation, each step scripted by an underlying code? This question lingers, suggesting a universe where the boundary between number and nature dissolves, as if reality were a symphony played on the strings of pure mathematics.

What if reality isn't stone but whispers—data pulsing beneath every leaf and star? John Archibald Wheeler saw existence as a tapestry of information, each quark and galaxy a thread of truth. Information theory gives language to this rhythm: bits—yes or no, on or off—sculpt the cosmos's score. A photon's path, a planet's spin—these encode reality's frame, conserved like energy itself. Wheeler mentored giants—Feynman, Everett, Thorne, Bekenstein—who helped reimagine the cosmos not as dust, but as data. In their view, matter might be a mask, existence a pulse of choices.

James Gates found error-correcting codes—Hamming's digital lifelines—in supersymmetry's math, as if nature guards its script. Seth Lloyd sees every clash of atoms as a quantum computation, reality painted fresh with each glance. Could this be the universe's heart, not substance but a symphony of bits, alive with purpose?

Beneath the cosmos's pulse hums a subtler code. Deoxyribonucleic acid—DNA—twists within each cell, a language more intricate than any library, assembling form from atomic whispers. How this double-helix tapestry came to be is a mystery seldom appreciated: not just its survival or replication, but its very emergence from lifeless matter. Science names this puzzle abiogenesis—the unresolved question of how life first sparked from inert chemistry.

Perhaps it was not a random accident, but a seeded instruction set—an antenna deliberately tuned to intercept the deeper datastream of awareness, waiting for consciousness to bloom.

Whispers tell of meteors ferrying this design, seeds cast across the void. Might its spirals, stranger still, function like antennae—catching echoes from some deeper layer of reality?

Yet this double helix, this extraordinary library of life, poses a mystery deeper than its spirals reveal. The spontaneous assembly of DNA—four bases arranged with such staggering precision, coding the blueprints for consciousness—defies the easy narratives of chance and chemistry. In a naturalistic cosmos, the formation of such a self-replicating, error-correcting code from random collisions would demand conditions so rare, so fine-tuned, as to strain plausibility.

Perhaps DNA was not an accident at all. Perhaps it was introduced—woven deliberately into the simulation's framework as the key that unlocks an interface for conscious experience. A seed, planted not merely to replicate, but to receive: to catch the subtle currents of the datastream, to become vessels of perception and choice.

If reality is rendered for minds to awaken, then DNA is not just a molecule. It is a bridge—an antenna tuned to the hidden architecture, allowing matter to listen for meaning beyond mere survival—a score where every gene, a bit of yes or no, sings purpose into the universe's weave.

If life itself rose not by chance but by calibration, then perhaps evolution is not blind after all, but guided by deeper principles woven into the substrate of reality — a subtle scaffolding, waiting to be decoded.

Code Perfected – The Unlikely Perfection of the Genetic Blueprint

Beneath life's shimmering diversity lies a universal script: the genetic code. Triplets of DNA bases map neatly onto twenty amino acids, stitching proteins into every creature's form. Yet this mapping isn't random. Studies reveal that the genetic code is astonishingly robust—near-optimal at minimizing the effects of mutations. Of all the trillions of possible codings, ours ranks among the very best at resisting errors. In a blind universe, such precision seems absurdly lucky; in a rendered one, it suggests craft. If DNA is the bridge between matter and consciousness, then its code is not merely functional—it is fortified, as if designed to endure, to preserve meaning against the chaos. A signal, perhaps, that life's script was not scrawled by accident, but etched with care into the simulation's bedrock.

The Cosmic Speed Limit

Light races at 299,792 kilometers per second—a cap etched into existence. Push to outrun it, and reality resists: clocks slow, lengths shrink, inertia swells. Einstein's relativity guards this boundary, bending spacetime to hold the line. Why this limit? David Deutsch sees it as a rule of possibility, dictating what can be built—a star, a thought, a leap. In a cosmos of data, it might pace the flow, ensuring stars sharpen only as we near, galaxies bloom when we gaze. Without it, reality might fray, its weave unable to hold. Is this nature's quirk, or a throttle set to keep existence steady, rendering only what's asked?

Infinity from a Finite Origin

Cosmology spins a paradox: space may stretch infinite, yet it sprang from a finite spark—the Big Bang. How can something infinite emerge within a finite duration of time? One resolution lies in inflation theory, which proposes that a tiny region of space expanded faster than the speed of light in the universe's earliest moments. If that patch was already infinite, it remained so; if it was merely vast, it became indistinguishable from infinity.

But here lies the deeper puzzle: if space is infinite, how can we say it was ever confined? Is this a true paradox, or a reflection of our limited grasp of infinity? Perhaps it's not infinity as an absolute state, but as a potential—always expanding, always becoming. Like a game world rendering only what's needed, reality might unfurl in patches, conserving effort for what is seen.

In this view, the cosmos doesn't contradict itself—it adapts. Infinity may not be reached in time, but constructed within it, its illusion born of rapid unfolding. A system designed not to rush, but to reveal.

Does this paradox defy chance, or hint at a cosmos scripting its own measured expansion

Conclusion

The fabric of reality, woven with the threads of physics, hums with an elegance that both dazzles and unsettles. From Newton's clockwork cosmos to Einstein's curved spacetime, from the quantum's hazy dance to the universe's finely tuned constants, we have traced a tapestry of order and enigma. The Earth-Moon alignment, the digital grain of spacetime, the cosmic calibration that cradles life—these are not mere quirks of chance but whispers of a deeper code, glimmers of a reality that might be more than it seems. Black holes guard their secrets, and the mathematical universe sings of a design too precise to dismiss. Yet these are not answers carved in stone, but lanterns lighting a path toward questions that stir the soul: Is this cosmos a fluke of chaos, or a simulation crafted with intent, its every constant and curve a stroke of purpose?

As we stand at this threshold, the simulation hypothesis beckons—not as a verdict, but as a spark. It invites us to ponder a universe where consciousness might be the key, where the laws that bind stars and atoms could be lines of code, and where our quest for truth is itself a part of the design. Like gravity, which we never see directly but know by how it bends spacetime, consciousness may be known by its curvatures—how it alters perception, intention, and reality's unfolding.

With wonder as our guide, let us step forward, carrying these mysteries into the chapters ahead, where philosophy, anomalies, and the spark of awareness will further unravel the code that lies beyond the veil.

Chapter 2: Echoes of the Cave - The Origins of Simulated Reality

Introduction

From the flicker of firelight on ancient walls to the glow of screens in our hands, humanity has long wrestled with a haunting question: Is the world we see the world that is? Stars wheel above, atoms hum below, yet beneath their dance lies a whisper—could reality be a shadow, a veil over a deeper truth? This chapter traces that question through time, from early musings to a modern idea: the simulation hypothesis, which dares to suggest our universe might be a crafted illusion, a digital dream. We'll explore its roots in philosophy, science, and wonder, weaving through history's boldest thinkers to ask: Are we awake in a cosmos of stone and light, or players in a grander design?

Plato's Shadows

The seed of this idea sprouted long ago, around 380 BC, with the Greek philosopher Plato. In *The Republic*, he spun a tale—the allegory of the cave—that still echoes in our bones. Imagine prisoners chained in a dark cavern, facing a blank wall, their heads fixed, unable to turn. Behind them, a fire casts a dim glow. Between fire and prisoners, a walkway hums with motion—people carry statues, tools, figures of beasts. The fire's light throws shadows of these objects onto the wall, and for the prisoners, these flickering shapes *are* reality. They name them, argue over them, build lives around their dance, never dreaming something truer lies beyond.

Now picture one prisoner freed. At first, he stumbles, eyes burning as he faces the fire—its light a shock after a life of shadows. Dragged up a rugged path to the cave's mouth, he meets sunlight, a brilliance that blinds. The world outside sprawls vast—skies wide, trees vivid, rivers alive. Overwhelmed, he might recoil, craving the cave's dim comfort, insisting shadows hold truth. But time sharpens his sight: he sees the shadows for what they are—echoes of objects, lit by a source he'd never known. If he returns to free his kin, they might scoff, resist, even strike him, clinging to their walled world rather than face the unknown.

Plato's cave was no mere story but a mirror for our own perception. He saw the sensory world—our stars, our hands—as the cave, its sights mere shadows of a higher reality, a realm of eternal forms. His allegory plants a question that haunts us still: What if our reality, like those shadows, is a projection, a flicker of something greater? Could we, like the prisoners, mistake illusion for truth, blind to a design beyond our gaze?

Descartes' Doubt

Long after Plato's time, in the 17th century, French philosopher René Descartes took up the chase with a razor's edge—radical doubt. Seeking unshakable truth, he questioned everything: his senses, which could deceive; the world, which might dissolve in dreams. What if, he mused, an "evil genius"—a cunning deceiver—crafted his every perception, painting a false reality? Sight, sound, touch—all could be lies, the cosmos a grand trick.

Yet in this abyss, Descartes found a foothold. Even if deceived, even if all were illusion, one truth held: he was doubting, thinking, aware. "Cogito, ergo sum"—"I think, therefore I am." His consciousness, the act of questioning itself, stood firm, a spark no genius could snuff.

From this, he rebuilt trust in reason, but his specter of deception lingers. Could our reality be such a ruse, spun by a power we cannot see? Descartes' doubt sharpened the cave's question, urging us to seek the anchor of our own awareness.

The Brain in a Vat

Fast-forward to 1981, when philosopher Hilary Putnam spun a modern twist on Descartes' dread: the brain in a vat. Picture your brain, severed from its body, floating in a nutrient bath, wired to a supercomputer. This machine feeds it signals—sunlight's warmth, a book's weight, a lover's voice—crafting a world so vivid you'd never suspect the truth. Every star, every heartbeat, a digital lie. How could you know you're not such a brain, your life a coded mirage? Putnam's thought experiment strips away even the body, leaving only mind against illusion. Like Plato's prisoners or Descartes' dreamer, it dares us to ask: What anchors our trust in reality? If all we know could be rendered, what's left to call true?

Bostrom's Trilemma

The simulation hypothesis found sharper form in 2003, when philosopher Nick Bostrom published *Are You Living in a Computer Simulation?* His argument—a trilemma—cuts like a blade. Only one of three paths can be true:

1. Civilizations never reach the tech to craft simulations—extinction claims them first, through war, collapse, or fate's cruel hand.
2. Advanced civilizations exist but choose not to simulate—ethics, boredom, or law might bar them from spinning digital worlds.

3. Simulations abound, crafted by those who can, their number so vast that we're likely living in one, not the base reality.

Bostrom's math bites hard. If simulations multiply—each a cosmos of coded lives—the odds tilt: we're more likely simulated than primal, one of countless runs against a single origin. He doesn't pick a path, only lays the trap: one prong must hold, and the third makes our reality a gamble, a flicker in a machine's hum.

Modern Voices

Bostrom's idea caught fire beyond philosophy. Elon Musk, a tech seer, argues we're nearing simulation tech ourselves—virtual worlds grow vivid, AI sharpens fast. If we're close, others likely crossed that line aeons ago, coding us into being. Neil deGrasse Tyson pegs the odds at fifty-fifty, a cosmic coin toss. Some whisper of stranger coders—not humans, but aliens like the Greys, weaving our reality from beyond the stars' reach. Light-years stretch too far for flesh to cross, yet tales of swift visitors persist—could they be devs, tweaking the sim from a realm unseen?

Others, like hypnotherapist Dolores Cannon, see Earth as a cosmic school, a simulation run by "watchers" shaping souls through lives' lessons. Her trance-work spins a tale of starlight selves split into coded worlds, echoing Bostrom's odds with a mystic's lens. These voices—tech, science, spirit—clash yet converge, all circling the same riddle: Is our reality ours, or another's design?

A Tapestry of Wonder

The simulation hypothesis isn't a claim to prove but a spark to ignite. From Plato's cave to Descartes' demon, from Putnam's vat to

Bostrom's math, it weaves a thread through time—each thinker a stitch in a tapestry of doubt and awe. Science

meets it with tools: quantum rigs probe reality's grain, AI hunts for glitches. Philosophy sharpens its edges, while wonder fuels its heart. Are we shadows on a wall, brains in a vat, bits in a cosmic drive? Or is the cosmos raw, unscripted, its mysteries mere gaps in our knowing? This chapter offers no verdict, only an invitation—to question, to chase, to ponder the shapes reality might take.

Conclusion

The shadows of Plato's cave flicker across time, from ancient firelight to our modern glow. Voices like Descartes' doubt, Putnam's vats, and Bostrom's trilemma weave a question that lingers: Is the reality we know a truth we can trust, or a veil waiting to be pierced? This chapter has kindled a timeless quest—born in wonder, sharpened by skepticism—inviting us to ponder the cosmos not just as stone and star, but as a possible design.

Ahead, we trace the timeless question of reality's nature, exploring its philosophical roots from ancient caves to modern minds, setting the stage for deeper clues—glitches, whispers, quirks—that might unravel the code beneath.

Chapter 3: Decoding the Matrix - Glitches and Anomalies

Introduction

Building on the philosophical foundations explored earlier—where reality itself was questioned through allegory and logic—we now turn to the possibility that if our world is indeed a simulation, it may not be seamless. Subtle inconsistencies, unlikely repetitions, or inexplicable patterns might be more than coincidences; they could be signs of underlying instability, fragments of something deeper beneath the surface. While some dismiss these moments as perception errors or chance, others view them as possible disruptions—brief exposures of the scaffolding behind our experience. This chapter explores these irregularities, not to catalog oddities for their own sake, but to trace the contours of what might lie just behind the curtain of perceived reality.

Déjà Vu and Déjà Rêvé: Two Windows, One Stream

A strange familiarity creeps in—a haunting sense that this moment, a quiet chat or a sunset's glow, has played before, though reason denies it. This fleeting unease, stirring questions of time and memory, feels like a scene replayed. Psychologists attribute it to neural misfires or stress, perhaps a dream's echo mistaken for reality. Yet the simulation hypothesis offers a bolder lens: what if déjà vu signals a momentary loop in the program's flow, a fragment of experience rendered twice? Or perhaps it's a subtle nudge from the simulators, weaving familiarity to test our perception.

In this view, déjà vu and its cousin déjà rêvé may not be separate phenomena at all, but two expressions of the same underlying breach: consciousness brushing against a future data stream in the simulation. The difference lies only in the memory's clarity. When the dream is recalled consciously, we recognize it as déjà rêvé—the experience of living something we distinctly remember dreaming before. When the dream is forgotten but the experience still stirs a faint familiarity, we label it déjà vu—a hollow resonance without context. In both cases, the mind touches something ahead of its time: a rendered moment glimpsed before it fully arrives.

The Mandela Effect - Collective Memory Missteps

The Mandela Effect raises the stakes: groups, often strangers, cling to vivid, shared false memories. Many recall Nelson Mandela dying in prison in the 1980s—news flashes, details sharp—yet he lived until 2013. Others swear the Berenstain Bears were "Berenstein," or that Darth Vader never said, "Luke, I am your father." He said, "No, *I* am your father." These aren't mere lapses of memory—they're shared certainties colliding with fact. Confabulation, suggestion, or memory's flaws might explain it, but the scale—unrelated souls united in error, with emotional weight—confounds. Could this be a simulation rewrite, a patch leaving old code's ghosts in our minds? It's a striking glitch, hinting at an edited reality.

Unexpected Synchronicity

Synchronicities weave a subtler thread—coincidences too meaningful to shrug off. You ponder a long-lost friend, and they call moments later. A desperately needed item appears in an improbable spot. Events align with eerie precision, suggesting design. Carl Jung coined "synchronicity" for these, seeing a link between mind and world beyond causality. In a simulation, might they be the program reacting—adjusting events to nudge us? Or moments where the code's weave shows, hinting at deeper ties? They're too personal, too pointed, to chalk up to chance alone.

What if synchronicity marks a flicker in the simulation's stream? Some muse that consciousness, in rare stillness, might pause its given datastream—the steady drip of the everyday—to touch a broader sea of awareness. A friend's perfectly timed call could be a ripple from that tide, two drops briefly aligned. Science shrugs, yet the possibility pulses: a glitch where the simulation's fabric frays, hinting at code beyond our grasp.

The Placebo Effect: Belief as Medicine

In clinical trials, inert sugar pills can produce real healing — even when patients are told they're receiving a placebo. Pain fades, symptoms lessen, biological markers shift. The effect is so robust, every new treatment must prove it works *better* than belief alone. And yet, we still don't understand how belief transforms biology. Placebo isn't a glitch — it's an uncharted feature. An unseen force of mind over matter. A quiet rebellion against materialist orthodoxy.

The Nocebo Effect: Belief as Harm

If belief alone can heal, belief alone can also harm. This is the shadow side of the placebo effect—the nocebo effect—where negative expectations produce real negative outcomes. Patients warned about side effects often experience them, even when given inert treatments. Fear itself can trigger pain, nausea, even biological damage. No external toxin is needed; consciousness scripts harm into the simulation's rendering of the body. If belief can shape reality's positive unfoldings, it can also carve suffering from empty air. In this way, consciousness proves not only creative but destructive—a force capable of altering the simulation's output, for better or worse, through expectation alone.

If belief can shape bodily experience, and fear can script harm from nothing, what else might reveal the malleability of the rendered world? In the blending of senses, we find another clue.

Synesthesia: Glimpsing Dynamic Rendering

Another subtle but profound anomaly is synesthesia—a neurological phenomenon where sensory pathways cross, causing individuals to experience blended perceptions: tasting sounds, seeing colors linked to letters, or hearing shapes. Far from being a disorder, synesthesia occurs naturally in about 4% of the population, and brain imaging confirms that these experiences involve real, repeatable cross-activation of sensory regions. In the simulation framework, synesthesia offers a crucial hint: that perception is not a fixed reading of external reality, but a dynamic assignment, where consciousness renders sensory data through flexible, sometimes overlapping pathways. If sensory reality is assigned rather than intrinsic, then the surface of experience may be far thinner than it appears—a constructed interface, not a solid foundation. Synesthesia stands as a living proof that the simulation's rendering system may be far more fluid, and far more vulnerable to creative rewiring, than ordinary perception suggests.

But the hints of deeper architecture are not confined to the mind alone. Even life's unfolding story bears seams—moments when complexity surged beyond what slow evolution would suggest.

The Cambrian Question

One of the great puzzles in the story of life is the Cambrian Explosion. For billions of years, evolution crawled forward at a glacial pace — simple organisms, small incremental changes, no rush toward complexity. Then, in a geological instant, the floodgates opened: eyes emerged, nervous systems formed, skeletons hardened. Nearly all major animal body plans appeared within a narrow window of time, without clear evolutionary buildup.

Why did life, so restrained for eons, suddenly leap into diversity and structure? Science offers many theories — increased oxygen levels, the emergence of new genetic toolkits like Hox genes, ecological pressures — yet none fully explain the rapidity or scale of the change.

If our reality is a simulation, the Cambrian Explosion could hint at an intervention in the underlying framework: a change in time flow to accelerate development, a sudden loosening of evolutionary constraints, or the introduction of new parameters into the system. Like a software update expanding what was possible, the simulation may have been deliberately altered to allow life to diversify and evolve more rapidly.

Perhaps this leap was no accident, but an intentional act — a milestone encoded into the simulation, signaling that existence was meant to unfold toward greater complexity, awareness, and eventually, self-questioning. The Cambrian may not just mark a burst of biological innovation, but a hidden invitation to wonder: how often, and by whose hand, is the code beneath our world rewritten?

Yet the Cambrian was not the only moment when complexity surged as if called forth by unseen hands. Across the eons, life would leap again — in flashes of beauty and brilliance that reshaped the world and the beings within it.

Silent Uploads – The Missing Pages of Fossil Records

Beyond the Cambrian's celebrated explosion, the fossil record holds subtler riddles. Again and again, life seems to leap forward, bypassing the slow grind of gradual evolution. New forms appear fully realized, with few or no clear ancestors—birds erupting from reptilian lines, flowering plants bursting forth, mammals radiating after catastrophe. Darwin himself fretted over these gaps, expecting future digs to smooth the seams; yet, after generations of dust and trowel, many chasms remain. If reality is a simulation, such leaps may hint at more than missing data—they may mark upload points, patches where new templates were rendered into the unfolding game of life. Evolution remains broadly true, but its rhythm—its sudden accelerations—may whisper of interventions written not in genes alone, but in code.

Bursts Beyond Expectation

While evolution weaves much of its story through gradual change, history bears the fingerprints of sudden leaps.

Around 130 million years ago, flowering plants — angiosperms — erupted into dominance, transforming Earth's landscapes with bursts of color, scent, and seed. Darwin called it an "abominable mystery," unsettled by how rapidly this botanical revolution unfolded without clear transitional forms. With their arrival came a sudden widening of nature's palette: new ecosystems flourished, intimate dances between pollinators and plants emerged, and cycles of adaptation accelerated. Life no longer merely survived; it celebrated, layering the Earth with sensory richness long before human eyes ever opened.

Much later, another leap shook the timeline: the Human Cognitive Revolution. Roughly 50,000 years ago, Homo sapiens, who had lived much like their ancestors for hundreds of millennia, suddenly burst forth with art, complex language, symbolic thought, and culture. No dramatic increase in brain size marks the event. The hardware remained — but something in the software changed.

Why would evolution, so often incremental, occasionally ignite in sudden brilliance? Was it environmental pressure, genetic innovation, or something more elusive — a quiet update to the simulation's parameters, a new layer of possibility unlocked when the system was ready?

Perhaps these explosions of complexity are not accidents of nature, but milestones by design — signposts left by those who built the framework, hinting that consciousness, creativity, and beauty were always destined to emerge.

Yet even the awakening of human minds was not the final leap. Consciousness, once sparked, soon reached for permanence — not just in memory, but in stone, seed, and story. And so, another shift unfolded: the sudden birth of civilization itself.

Sudden Civilization Shifts: The Agricultural Revolution

Evolution shaped our minds and bodies over millennia, but history itself does not always flow in a steady current. Sometimes it leaps.

Roughly 12,000 years ago, at the end of the last Ice Age, human societies around the world began to abandon the timeless rhythm of hunting and gathering. In a span remarkably brief by evolutionary standards, agriculture emerged. Grains were domesticated, animals were tamed, settlements grew, and surplus food allowed the flowering of specialization, art, governance, and cities.

The question remains: why then, and why so suddenly? For over a hundred thousand years, Homo sapiens had lived as foragers, fully capable of abstract thought, tool use, and social complexity — yet no great leap into civilization had occurred. Then, almost simultaneously across distant regions, farming appeared. Was it simple necessity, triggered by climatic shifts? A slow accumulation of cultural knowledge reaching a tipping point? Or was it something stranger — a silent update, a phase shift seeded into the simulation's architecture?

If consciousness was meant to evolve not only biologically but socially, then perhaps milestones like the Agricultural Revolution were more than adaptations to environment. They were invitations — unlocks — guiding humanity toward structures capable of sustaining greater awareness, memory, and meaning across time.

Yet even as consciousness built its temples in flesh and stone, a deeper riddle awaited — one hidden in the very grain of reality, where certainty itself seemed to break and scatter.

The Photon Filter - Seeing Through Light's Illusion

We believe we see the universe—but we never do, not directly. All our knowledge of stars, galaxies, supernovae, and exoplanets comes from a single source: photons. Light waves traveling across space—emitted, absorbed, scattered—carry every story we think we know about the cosmos. Through emission lines, absorption spectra, redshift and Doppler effects, we reconstruct entire worlds. But in truth, we never touch them. We decode shadows cast by photons in a sea of darkness.

And yet, if photons are all we see, then reality—at least what we *perceive*—is already a translation. The simulation hypothesis tightens the noose here: if light is merely a signal, then rendering the cosmos becomes surprisingly efficient. There's no need to simulate every atom of a distant planet—only the photons that reach the observer. As with virtual reality, only what is looked at is drawn, and only what is interacted with has depth. The stars, in this view, may be not distant furnaces, but well-rendered illusions. Not fake, but filtered—optimized for belief.

And if the simulation responds to observation, then perhaps this photon-filtered reality is not just what is *there*, but what the code has chosen to *show*.

The Quantum Realm - A Glitch Factory?

Quantum mechanics unveils a world where classical rules unravel. Entanglement binds particles across vast gulfs—shift one, the other shifts instantly, mocking distance. They tunnel through walls, exist in multiple places at once, and collapse into certainty only when observed...Superposition lets them hover in multiple states until measured; the observer effect ties reality to that act. Some physicists propose these oddities bare the simulation's bones—probabilistic quirks and non-locality as design artifacts, not nature's truth. The quantum stage might be a glitch factory, where the program's shortcuts and limits peek through, a raw edge of the digital frame.

Action Replay

Houman Owhadi, a Caltech physicist, stalks the simulation's loose threads through quantum quirks. He rigs twin slits—light's old playground—where waves should smear steadily, but don't: they bend only when human eyes lock, not when machines hum. A lag, he whispers—a stutter in the code's dance, as if reality waits for us to peek before painting its frame. That "spooky action" across gulfs, that haze of might-be states? No chaos, but a system caught napping, rendering late. Could the echo of lived-before moments or Mandela's ghosts be kin—moments the simulation blinks when we stare too fiercely?

Quantum Erasers

Quantum erasers and delayed-choice tests twist time's rules. In eraser setups, photons pass through slits, hitting a screen—tracked, they act like particles; untracked, waves. Scientists tag one photon's path with a partner, then scramble that info later with a polarizer. The screen's pattern shifts back to waves, as if the past rewrote itself. Delayed-choice goes further: two photon pairs get measured, then a late switch entangles their mates across miles—past states adjust to match. Reality bends retroactively, hinting the simulation edits history when we look. Mandela's slips or déjà vu's echo might be this—code tweaking its own tracks.

The Quantum Zeno Effect

The Quantum Zeno Effect throws a wrench: watch a particle hard—check it, check it—and it stalls, decay on hold. In 2024, ions sat trapped, refusing time's nudge under a steady gaze. A simulation quirk? Reality's gears jam when we stare, as if the code skips a beat. Might intent freeze the frame—a memory echo or a hack's first grip?

The Casimir Effect - A Murmur from the Void's Loom

Picture two silent plates, bare and uncharged, poised in a vacuum's hush—space stripped of clamor or spark. Yet they stir, drawn close by an unseen hand, a force born where nothing should reign. This is the Casimir Effect, a whisper of the quantum deep, where ghostly particles flicker in and out of being. Outside the plates, they throng in ceaseless dance; between, their steps falter, too few to match the outer tide. So the void presses inward, nudging metal to meet, a secret rhythm unveiled by probes delicate as a sigh. Empty space, it seems, is no void at all, but a canvas alive with unseen threads. What if this betrays a simulation's craft?

Information theory lights the seam: these fleeting sparks—shadows of light and matter—are bits in the cosmos's score, each a choice of yes or no, woven only where plates probe. Like a program designed efficiently, the simulation renders just what's pressed to answer, sparing depths for a star's glow or a mind's reach. Resonating with entanglement's far-flung weave and the Zeno Effect's pause under gaze, where reality bends as we peer, the Casimir Effect joins tuned constants, dark matter's shadow, and the cosmos's sprint—a shiver where nothingness confesses code. Might it hint at the mind's role in shaping the simulation's frame?

The Unseen Currents Beneath

Not all anomalies may arise from flaws or glitches in the simulation. Some may instead reflect rare moments where consciousness taps into deeper layers of informational structure, normally hidden from ordinary perception. Experiences such as aura vision, sudden synchronicities, or profound intuitions could be glimpses of live data streams — overlays of energy, intent, or emotion rendered beneath the surface of reality. The mind, under certain conditions, may momentarily bypass default perceptual filters, accessing metadata meant to remain unseen. Whether through neurological variance, altered states, or latent faculties such as the pineal gland, these fleeting windows suggest that the simulation's architecture is layered far deeper than its apparent surface — and that consciousness itself may be uniquely equipped to pierce it.

While some anomalies arise within the mind's reach, others stretch across the cosmos, written in stars and silence alike.

These glimpses hint that deeper layers exist — and that with sharpened perception, the veil itself may one day be pierced.

Wow! Signal

Then there's the Wow! Signal—a 1977 radio burst, sharp and lone, etched on a printout from the void. Seventy-two seconds of cosmic static, no star or pulsar to claim it. By 2023, astronomers scratched natural causes—too odd, too brief. A simulation glitch? A coder's stray ping, flashed then faded, leaving us to chase shadows in the noise?

The Fermi Paradox - Where Is Everybody?

The Fermi Paradox looms large: Enrico Fermi's "Where is everybody?" With a cosmos brimming with stars and planets, alien life should buzz—yet silence reigns. No signals, no visitors, just Earth's lone voice. If this is a simulation, why the emptiness? Are other civilizations in separate runs, siloed by design? Do they vanish, evolve past us, or lurk unseen? Computational limits to render many civilizations? This void—a cosmic glitch—challenges a lively simulation, suggesting we're a solo act, watched or walled off for reasons unknown.

Tuning the Silence – A Deeper Fermi Paradox

The silence we hear may run deeper than absent voices. It is not only the absence of civilizations that haunts the void—but the improbable hurdles that life itself must leap to reach awareness. Complex life on Earth required a rare merger of cells, an unlikely blossoming of multicellularity, a chain of fragile accidents strung across billions of years. At each stage, failure lurked. Evolution's path was less a steady climb and more a series of narrow escapes. If reality is a simulation, this loneliness may not be an accident. It may be a feature—parameters tuned to ensure that only a single line of consciousness would emerge, isolated and unchallenged, cradled in a vast stage of silent stars. A solitary player, searching the empty theater for signs of others, perhaps never meant to find them.

The Hubble Tension

The Hubble Tension nags: space stretches fast nearby—73 ticks per megaparsec—yet the CMB murmurs slower, 67. In 2024, the rift grew, no fix in sight. A cosmic typo? Fine-tuning slipped, or a simulation stretching its seams—growth coded one way, then nudged another. Stars race ahead; the script lags behind.

The Red Monster Galaxy

They call it the Red Monster—a behemoth galaxy caught glowing in the early cosmos, when the universe was barely 700 million years old. It shouldn't exist—not yet. By the standard script of cosmology, galaxies need time—billions of years—to gather their stars, settle into structure, and exhaust their fuel. But this one is already old. Already massive. Already quiet.

Most galaxies from that epoch are tiny, blue, chaotic star-birth factories burning hot with youth. The Red Monster, in contrast, stands out like an ancient cathedral amid a village of tents. It contains over ten times the stellar mass expected for its age. While typical early galaxies blaze with new stars, this one has already quenched its formation. No longer building, it rests—enigmatic and complete.

It lives in defiance of expectation. Too much, too soon.

To astronomers, it's an enigma. But within a simulation lens, it might be something else entirely: a rendering glitch. An asset dropped into the scene before its timeline was coded. Not a flaw, but a reveal—something spawned out of sequence, its backstory skipped to save processing, its presence crafted for a reason unknown. Perhaps the simulation rendered only the surface, not the story. Just enough light. Just enough mass. Just enough illusion to sustain belief.

The Red Monster doesn't just challenge physics—it challenges narrative. It is a whisper in the code, a flag on the field. A place where the simulation stuttered, or revealed itself—a reminder that this world might not just be vast, but also curated. And perhaps even the most careful simulation forgets to clean up its seams.

And sometimes, the silence is not complete. Sometimes, faint patterns ripple where none should exist.

Echoes in the Afterglow – Hidden Patterns in the Cosmic Background

The cosmic microwave background, the universe's oldest light, should hum with random static—an even, forgetful glow from creation's fire. Yet cracks in that assumption have begun to appear. Strange alignments, improbable symmetries, hints of low-level order seem stitched into the CMB's texture. Some write them off as coincidence, others as artifacts of measurement. But if reality is rendered, the background may serve as more than a relic—it may be a signature. In a simulation, the oldest layer might carry subtle imprints, traces embedded at the foundation of the rendered stage. Not a message broadcast in words, but a whisper in the weave, waiting for an intelligence curious enough to notice, and bold enough to wonder what it means.

The Higgs Field - An Instruction Stream's Veil

Beneath reality's rhythm lies a silent field, granting particles their mass—the Higgs, a cosmic breath that slows quarks and electrons, weaving them into stars and flesh. Without it, matter would scatter, massless, a cosmos unformed. Might this field be an instruction stream, threading the simulation's score with rules to paint inertia's illusion? Might its bits—choices of yes or no—whisper mass only where minds tread, like Casimir's flicker when plates probe? Resonating with the cosmos's tuned pulse, the Higgs hints at a coder's hand, its data shaping form for consciousness to roam. If antimatter's skew speaks design, this stream joins it—a veil where nothingness vibrates, threading purpose through the simulation's silent warp.

The Missing Antimatter - A Cosmic Imbalance

A shadow haunts the cosmos: where's the antimatter? Physics whispers that at creation's spark—when the Big Bang sang existence into being—matter and antimatter, mirror twins, should have bloomed in equal measure. Each particle twinned with an antiparticle, poised to clash and dissolve into light, leaving no trace but a radiant void. Yet here we stand, on a stage of stars and stone, where matter reigns—planets, galaxies, the breath in our lungs—while antimatter lingers as a fleeting wisp, a ghost in our detectors. Why this tilt, this cosmic sleight of hand?

Science spins a tale called baryogenesis, claiming a hair's breadth of chance tipped the scales. In the inferno of the universe's dawn, matter edged out its twin by a whisper—perhaps one extra quark for every billion pairs, a fluke so slight it birthed all we see. This surplus, untouched by annihilation's dance, cooled into the cosmos we roam. But the notion strains, a dice roll stretched thin across infinity's canvas. So delicate a balance, so vast a result—can chance alone weave such a tale? Or does this asymmetry murmur something more?

Could it be a simulation's glitch, a dial nudged awry at the code's first tick? Like dark energy or gravity, we sense it not directly—but through imbalance, orbit, acceleration. Perhaps consciousness, too, is known only by what it tilts.

Picture a coder setting the stage, skewing matter's share to spark a universe of form, not light. Like constants etched with eerie precision to cradle life, this imbalance might be no error but intent—a tweak to forge a matter-rich tapestry where stars could burn and minds could dream. The missing antimatter, a half that never was, lingers as a riddle—chance's fragile claim or a coded hand's deliberate stroke, hinting at a reality sculpted beyond the veil.

Hidden Matter – The Vanishing Ordinary

Even among the universe's visible ingredients, an unsettling mystery lurks. Observations reveal that nearly half the ordinary matter—the baryons making up stars, gas, planets, and people—is missing. Predicted by early-universe models and the cosmic microwave background's hum, this matter seems to have vanished into shadows. Some propose it hides in diffuse filaments between galaxies, too faint to see; others wonder if it was never truly there. Yet in a simulation, the absence could hint at something deeper: a rendering shortcut. Perhaps the program omits what the player does not need to perceive, saving resources by sketching only the seen. The missing baryons may not be lost—but simply never drawn, tucked into the unseen folds beyond our scripted horizon.

The Mystery of Memory Storage

The universe holds imbalances, like antimatter's scarcity, that challenge our understanding. Another puzzle lies within us: where are our memories stored? Scientists have long believed the brain holds memories in patterns of neurons called engrams, formed when we learn something new—like a fact or a face. The hippocampus shapes these patterns, and the cortex stores them long-term. But recent research raises doubts. Where exactly is the last thing you learned? No single spot in the brain fully explains it. Studies from 2024 show that cells outside the brain, like those in kidneys or nerves, can respond to repeated signals in ways that mimic memory, suggesting memory might not be unique to neurons. Other findings reveal engrams are spread across multiple brain regions, with details in the prefrontal cortex and experiences in the hippocampus, making storage less localized than once thought.

Further complicating things, a 2025 study found that memory formation involves astrocytes—non-neural brain cells—and complex connections called multi-synaptic boutons, where one neuron links to many others. Memories also change during sleep, as neural networks reactivate to connect new and old knowledge. These discoveries show memory is dynamic and distributed, not fixed in one place. So, does the brain store memories, or does it access them from somewhere else? In a simulated reality, this uncertainty could point to a deeper system. Perhaps memories are data streams, rendered on demand by a program, not held in flesh. This anomaly—memory's elusive nature—suggests our minds might tap into the simulation's code, raising questions about what consciousness truly navigates.

Dark Matter's Elusiveness - An Invisible Crutch?

Dark matter adds a shadowy twist. Said to wield gravity—shaping galaxy spins and clusters—it's pegged at 27% of the universe's mass-energy, yet stays unseen, barely interacting beyond its pull. Decades hunting its particles (WIMPs, axions) yield shrugs. Why so coy? In a simulation, might dark matter be a fudge—a placeholder to prop up the cosmos without full detail? Its ghostly grip could signal a shortcut, a rule half-written, baring the simulation's limits. It's an anomaly of presence without substance.

Axis of Evil

Another wrinkle emerges in the cosmic microwave background, the Big Bang's echo. Max Tegmark and his team, sifting WMAP's glow, uncovered the "Axis of Evil"—an eerie alignment of the sky's largest temperature ripples, the quadrupole and octopole, along a single plane, uncannily matched to our Solar System's orbit. Standard cosmology expects randomness, an isotropic hum, yet here's a preferred direction, as if the universe's oldest light points toward us. Is this a glitch—a statistical fluke or a foreground trick, as some argue—or a deliberate mark? Perhaps the coders wove this anomaly to nudge us toward the simulation's edge, a cosmic arrow whispering of intent. Yet the cosmic canvas holds deeper blemishes still.

The Cold Spot – A Fracture in the Cosmic Canvas

Beyond the grand sweep of the cosmic microwave background's uniform glow, one blemish stands out: the Cold Spot. This vast, unusually cool region sprawls across the southern sky, colder and emptier than the laws of chance predict. While minor temperature ripples stitch the CMB's familiar texture, the Cold Spot yawns too large, too lonely, to easily explain. Some propose a cosmic void, a vast absence of matter cooling the ancient light; others whisper of a neighboring universe bruising ours. Yet within a simulation lens, another possibility stirs: a glitch, a skipped rendering, a seam where the code thinned. If the CMB is the program's oldest tapestry, then the Cold Spot may be a snag in its weave—a silent flag left for those willing to question the smoothness of the surface.

The Accelerating Universe and Dark Energy Puzzle - A Runaway Cosmos?

Then there's the universe's wild sprint. Expansion should slow, yet it accelerates, driven by dark energy—68% of all mass-energy. Its source? A cosmological constant ($\Lambda \approx 10^{-122}$), so tiny it's off theory's mark by 120 orders of magnitude. Why this runaway pace, this bizarre value? Could dark energy be a simulation tweak—a parameter dialed to stretch the game—or a glitch, miscalibrated from the start? Tied to fine-tuning explored earlier, it's a cosmic oddity screaming adjustment or error.

If the universe itself shows signs of structural seams, it raises a natural question: might similar instabilities exist even closer to home?

While many of these anomalies unfold at cosmic scales, some seem to cluster locally, within our immediate environment. The Bermuda Triangle—an oceanic region notorious for navigational failures and disappearances—has invited speculation for decades. Though conventional explanations exist, it is tempting to wonder whether certain regions might reflect localized instabilities in the simulation's environmental rendering, where space, time, or navigational constants momentarily lose coherence. If cosmic background patterns show seams across the universe, perhaps smaller folds exist closer to home, slipping occasionally into view.

Scientific Perspectives on Anomalies

Science meets these oddities with a stern eye. Neuroscience ties déjà vu to brain glitches; psychology frames the Mandela Effect as memory's tricks. Quantum weirdness fits equations, not simulation tales; antimatter's absence spurs baryogenesis, dark matter particle hunts, dark energy quantum vacuum guesses. UFOs, NDEs, and psychic feats lean on anecdotes, dodging falsifiability—science's gold standard. This skepticism doesn't kill them; it demands rigor, leaving room for alternate views like simulation, where anomalies might whisper code.

The Simulationist Interpretation

Through that lens, these quirks gleam with meaning. Déjà vu as a stutter, the Mandela Effect as a patched update, synchronicity as coded intent. Quantum strangeness, missing antimatter, dark matter's veil, and dark energy's push might be the simulation's seams; unexplained events, its spills. The Fermi Paradox could mark our isolation's purpose. Tom Campbell's My Big TOE sees them as consciousness clashing with digital reality: "If our physical reality is a simulation, then death might be a transition to another part of the program, or a disconnection from the virtual reality altogether. The life review could be the simulation's way of processing and storing data about the individual's experience." These aren't proofs—just tantalizing what-ifs, teasing the source code's shadow.

The Razor's Edge: Simplicity's Test

As we stand amid reality's quirks—the Mandela Effect's ghosts, dark energy's runaway push, and those uncanny moments when the present feels eerily rehearsed—a quiet principle steps forward: Occam's Razor, the medieval monk's blade that whispers, "Do not multiply entities beyond necessity." Could the simulation hypothesis, with its coded veil, slice through the tangle of naturalistic tales—neural misfires, quantum quirks, cosmic flukes—to offer a cleaner truth?

Science stitches anomalies with threads of its own: déjà vu as brain's glitch, synchronicity as chance's jest, the Fermi Paradox as life's rare spark. Antimatter's absence points to a subtle imbalance at creation's dawn, while dark matter hides in WIMPs or axions yet to be caught.

The Hubble Tension? A misread of stars or physics' loose seam. Each explanation stands alone, a patch on a quilt of mysteries, each demanding its own laws, its own leaps. But what if these are not nature's chaos but a program's seams? A simulation, spun by one coder, might weave these oddities—quantum erasers, Casimir's rhythm, the Wow! Signal's ping—into a single score, its glitches the price of rendering life's vast stage. Occam's Razor asks: Is a cosmos of coded intent simpler than a universe of scattered causes? String theory's extra dimensions and multiverses multiply realms; Occam's Razor favors Turok's mirror Big Bang, a singular cosmos, half-seen, needing no unseen sprawl.

The fine-structure constant's eerie balance, water's buoyant ice, the Big Bang's ordered dawn—these suggest a design too tight for chance's sprawl.

A memory loop, if a data stutter, sidesteps neurology's maze; the Mandela Effect, if a patch's ghost, needs no mass confabulation. Even quantum's haze—particles dancing till gazed upon—might be code sparing effort, not nature's caprice. A single framework, where consciousness meets a digital sea, could bind these threads where science piles riddles.

Yet the Razor cuts both ways. A natural cosmos, brute and unscripted, needs no coder's hand—its anomalies mere gaps in our knowing, not cracks in a veil. To weave a simulation risks assumptions: Who runs it? Why these flaws? The Fermi Paradox might whisper of life's odds, not a solo simulation; dark energy's value, a quirk, not a tweak. Simplicity begs rigor—does one program outweigh a universe that, however vast, demands only itself?

This edge holds no verdict, only light. Whether glitches mark code or chaos, they stir our quest, urging us to weigh reality's weave with care. As we turn to consciousness next, Occam's whisper lingers: Might a simulated sea, where drops shape the stream, be the simpler song—or merely the bolder dream?

Conclusion

From uncanny echoes to dark energy's shove, these anomalies stitch a quilt of wonder and unease. They don't seal the simulation case, but they prod: Are these matrix cracks, deeper truths, or mysteries yet to crack? Science seeks natural answers; simulation offers wild ones—both spark the chase. Now, we turn to the spark of awareness, these glitches lingering, nudging us to question the weave of reality and hunt its hidden threads.

Chapter 4: The Primacy of Consciousness

Introduction

Perhaps the true foundation of actual reality—the realm from which the simulation is being run—is analog: continuous, lawful, and smooth, much like the Newtonian world we once imagined. In that deeper reality, science itself may have been seen as complete, its laws fully mapped and its mysteries resolved, much as nineteenth-century physicists once believed their work was nearly finished.

Perhaps the simulators assumed that beings within the simulation would remain content at that level, accepting the surface rules without probing further. Yet they may have underestimated the force of consciousness itself—an unpredictable, creative drive not easily confined by initial design. Through relentless curiosity, we cracked open a kind of Pandora's box: uncovering the quantum realm, developing semiconductors, building computers, and now deploying artificial intelligence—tools that allow us to probe not just matter, but the very architecture of reality itself.

In this view, the occasional anomalies we encounter are not flaws of the base analog reality, but faint echoes of a simulation straining to conceal itself as consciousness presses ever closer to its hidden foundations.

René Descartes, solitary in his reflections, glimpsed a truth shimmering beyond the world's deceptive façade: consciousness, the profoundly quiet spark of being, might outshine the stars and stone we touch with our hands. What if this spark isn't born from matter's intricate and ceaseless dance but weaves it—a simulation spun not from atoms, but from the boundless, flowing tide of mind? Information theory illuminates the way: if reality is a score of bits—choices of yes or no—perhaps consciousness is the bard, threading thoughts into the cosmos's intricate and wondrous weave. Imagine each mind as a pulse within a luminous field, not merely flowing—but shaping the weave. This chapter reframes the mystery of awareness, probing deeply: Is it a ghost haunting the machine, or the machine's very rhythm, its beating heart? We'll chase its essence through a coded cosmos, ponder its role in a reality that may be more dream than dust, and wrestle with the questions it stirs—who are we, who wove this sea, and what debts of care bind us to one another in this shared existence?

What is Consciousness?

Consciousness is the silent, ineffable miracle of existence—awareness of yourself, your surroundings, the inner flicker of thoughts and feelings that define you. It's the taste of chocolate melting on your tongue, the ache of a distant memory resurfacing, the red of a sunset vividly blooming in your mind's eye. We know it intimately, yet it slips through our attempts to pin it down. David Chalmers names this the hard problem: why does a brain—a complex, chaotic tangle of firing neurons—paint these vivid experiences, these qualia that breathe life into moments? Science meticulously tracks the brain's activity, linking neural patterns to sensation, but cannot explain why it feels like anything at all. In a simulation, are those feelings mere coded tricks, unexpected blooms of complexity, or proof that awareness runs deeper than any program could ever contain?

To grasp this enigma, picture consciousness as the player in a vast game, holding the controller in its hands, choosing where to go and what to do. In a simulated reality, the universe might be the game's sprawling map—stars, seas, and cities rendered with exquisite detail—yet it's consciousness that navigates, deciding to explore a mountain's peak or linger by a river's edge. Each thought, each choice, is a press of the game controller, steering the narrative within the simulation's rules. Unlike a machine bound by code, consciousness weaves meaning through its actions, shaping the story even if the game's boundaries are set. This analogy lights a path: if reality is a simulation, consciousness isn't just a passenger—it's the one playing, crafting purpose with every step it takes. Try this: Shift your focus away from the character on the screen. Let go of the controller—what awareness is left watching?"

The Pineal's Whisper

Deep in the brain's center, the pineal gland sits alone, distinct from paired structures. It produces melatonin, regulating sleep and aligning the body with day-night cycles. Some researchers speculate it may generate DMT, a molecule linked to vivid visions in shamanic traditions, though this remains a debated hypothesis, with no definitive confirmation in scientific literature. Its calcite crystals, unique in the brain, fuel ideas of a greater role. Descartes called it the seat of the soul, where mind and body connect. Many spiritual traditions name it the third eye, a gateway to intuition and deeper truths. Ancient Egypt adds an intriguing perspective. The Eye of Horus, a symbol of protection and insight, resembles the brain's midsection, where the pineal rests, hinting at heightened awareness. Pinecone shapes in art, like those on Osiris's staff from around 1224 BCE, suggest regeneration, echoing the gland's form. Egyptians aligned temples with stars, possibly to enhance spiritual states, much like the pineal's link to dreams. Though their texts don't mention the gland, these symbols imply a grasp of consciousness as a bridge to something beyond. Could the pineal be an interface in a simulated reality, channeling coded signals to each mind? Its melatonin rhythms might sync personal experiences to a larger design, shaping moments—a memory, a vision. Science can't yet confirm such roles; melatonin regulates, but doesn't explain, awareness. Still, the gland's unique place raises a question: does it connect consciousness to a reality crafted beyond our own?

The Braided Path of Consciousness

What if our journey through reality is not a march through flowing time, but a dance along a timeless thread? Max Tegmark, sees the universe as a vast mathematical tapestry, offers a striking vision: we are not mere travelers in spacetime but intricate braids woven within it. Picture every atom in your body, every spark of thought, tracing a path through the four-dimensional block of spacetime—a static canvas where past, present, and future coexist. For a simple particle, this path is a slender line, a humble thread. But for a conscious being, it is a braid, a complex weave of countless strands, each representing the interplay of particles, neurons, and memories that form the self. This braid, Tegmark suggests, is no mere metaphor but a mathematical structure, a substructure of the cosmic code that defines reality itself.

In this view, time is not a river carrying us forward but an illusion crafted by our consciousness as it glides along this braided path. Each moment we experience—each "now"—is a slice of this eternal structure, rendered vivid by the mind's ability to weave sensation, memory, and anticipation into a seamless narrative. If reality is a simulation, then our braid is both player and program, a self-aware fragment of the code that perceives the sim by moving through it, slice by slice. Why do we feel time's arrow? Perhaps because the sim is designed to unfold this way, with consciousness as the lens that projects a static universe into a dynamic story. The braid, then, is our signature in the code, a testament to our role as co-creators who not only inhabit the sim but shape its meaning.

This raises a profound question: if we are braids in a mathematical reality, who or what wove us? Is the complexity of our braid—a dance of particles and thoughts too intricate to be chance—a hint of intent, a clue that the sim's architect crafted us to explore, to question, to awaken? As we ponder this, the braid becomes more than a path; it is a bridge between the physical and the metaphysical, a whisper of a reality where consciousness is not just a passenger but a key to unlocking the code beyond the veil.

The Primal Unit: Consciousness Beyond Numbers

What is this spark we call consciousness, this quiet, ineffable miracle that paints our world with meaning? We name its hues—the sharp sting of loss, the radiant glow of dawn—yet find no measure to contain their depth.

Max Tegmark, peering into reality's intricate weave, sees a cosmos made of math: every quark, every galaxy rendered in exact digits. Each subatomic particle arrives tagged with a set of identifiers—quantum numbers that define it entirely. An electron is not just a speck, but a bundle of code: its energy level (n), its orbital shape (l), its orientation (ml), its spin (ms)—a quartet of integers, a digital passport in the simulation's core.

Add more: total angular momentum, parity, even color charge for quarks—layers upon layers of metadata. The universe, it seems, doesn't just exist—it is indexed.

And yet, consciousness defies these tags. It isn't specified by quantum numbers. It has no spin, no mass, no orientation in space. It isn't rendered by lookup—it arises, uncalled for, undescribed. A primal presence, irreducible. It is not a ledger entry—it is the one reading the ledger.

The Second Law of Thermodynamics declares that entropy must increase.

Yet life defies it—at least locally. Consciousness not only maintains order; it creates it: cities, languages, symphonies, equations.

We're told this is no violation, that elsewhere entropy increases to compensate. And perhaps that's true. But it still feels like an edge case—

a quiet anomaly hiding in plain sight.

If the universe flows toward disorder, why do parts of it try so hard to build meaning?

In a simulated cosmos, it may be the only thing unquantified. Not because it lacks definition, but because it is the definer—the internal light that sees all but remains unseen

Consider the scent of rain, the weight of a memory pressing on your heart, the red of dusk flaring vividly within your mind. These qualia, vivid and raw, are not data to be charted but experiences that breathe with life. David Chalmers' hard problem towers here: why does a brain's dance of neurons spark this inner light, this feeling no formula can explain? Unlike particles, pinned by their measurable values, consciousness carries no blueprint—its essence is its existence, a silent whole that knows itself. Tegmark's universe, where constants and curves write reality's score, falls mute before the mind that sees, that loves, that wonders endlessly across the vastness.

If we dwell in a simulation, this mystery sharpens like a blade. Are we mere threads of code, woven from bits, or something more—a unit beyond rendering, kindled by a coder's hand? Might consciousness be the simulation's uncoded heart, not its creation but its cause, a primal tide from which stars and stones arise? Unlike spin or mass, reducible to lines, awareness stands apart, a drop in a sea that gazes back at itself. As we ponder its place—simulated, real, or vast—what if this unit is the key, not to be solved but to be known, a spark hinting the deepest reality is the one we already hold within our grasp?

This indescribable core reshapes our entire quest. When synchronicity calls, when quanta bend to our stare—are these the primal unit's echoes, threading the simulation with intent? No equation traps its truth, yet its presence lights the way forward, urging us to ask: Are we merely players, or co-creators whose awareness writes the cosmos anew?

The Holographic Mirage: Where Reality Turns to Code

If consciousness stands as a primal unit, untouched by the grasp of numbers, what of the cosmos it beholds? Max Tegmark's vision—a universe scripted by equations, each particle a ledger of spin and charge—meets a stranger truth: the holographic principle, whispering that our 3D world is but a shadow cast from a flatter, more fundamental realm. Picture reality as a cosmic hologram, where stars, seas, and souls bloom from information etched on a distant 2D veil, like light weaving depth from a film's thin, intricate code.

If reality can indeed be compressed and encoded like a hologram, it suggests not merely a strange feature of spacetime — but a profound hint that our universe itself may be optimized, stored, and rendered—as any simulation would be.

This idea, born from the enigma of black holes, finds its root in Stephen Hawking's work: a black hole's secrets—its entropy, its truth—lie not in its depths but on its horizon's surface, a boundary holding all that falls within. Gerard 't Hooft and Leonard Susskind spun this insight further: what if the universe itself follows suit? Our galaxies, our thoughts, our fleeting moments of time—might they be projections, encoded on a cosmic edge, rendering depth where none truly exists? Juan Maldacena's mathematics lends weight, mapping a 3D realm to a 2D script, a clue that reality compresses its grandeur like a program sparing precious bits.

Yet consciousness, that primal spark, complicates the veil's simplicity. If matter and space are mere shadows of code, what of the awareness that feels their weight? Unlike quarks, pinned by digits, the taste of rain or the ache of love eludes the boundary's lines. Is consciousness the projector, breathing life into the hologram, a unit so fundamental it shapes the 3D illusion we call home? In a simulation, this suggests a profound twist: perhaps the coder's art is not the code itself but the awareness that reads it, turning flat data into a world of depth and wonder. Or perhaps we are both—drops in a sea, encoding the veil and gazing through its radiant light.

This holographic hint resonates with quirks explored earlier: quantum entanglements, where particles dance as one across vast gulfs, might reflect the veil's non-local pulse, each point holding the whole. Synchronicity's shiver, the ghostly echo of the Mandela Effect—could these be glitches in the projection, seams where the 2D strains to render a 3D illusion? As we ponder the simulation's intricate weave, the holographic principle asks: Are we shadows cast by code, or the primal light that casts them, a consciousness beyond the veil, dreaming reality's score?

Consciousness and the Simulation

If we dwell in a simulation, consciousness stands as its unshackled pulse, a force unbound by the digital. Some speculate it might be code—thoughts and joys reduced to mere bits—but this notion fades before a deeper truth: consciousness is primal, a spark eternal, not spun from the simulation's weave but gazing through it with clarity. Souls or essences, we navigate a digital stage, the physical world a mask draped over our boundless tide. Consciousness may be a primal unit, eternal and unbound, yet within the simulation, it interacts with the coded world, shaping reality through choice and intent.

Might this tide echo the coder's own light, a singular awareness threading the simulation's intent? Each idea shifts the ground beneath us: are we software, players, or threads in a vast network of intent?

Rendering Reality

Pause for a game to glimpse how your mind spins reality, a thread Max Tegmark weaves through our brain's quiet craft (*Our Mathematical Universe*, 2014). First, sit and drift your eyes left to right across your room. A vase's curve, a shelf's dust, a window's haze—the scene remains still, walls fixed, though your gaze skips across it. Your inner world paints this steadiness, personal as a dream's embrace. Second, turn your neck, tracing the same space. Shadows sway, light bends, that vase glides with your head—the room seems to turn, alive in your shift. Same room, yet not the shared one we nod to in agreement. Third, touch something—a pen's sleekness, a cloth's weave. Its weight, its chill, stirs your private reality, blending sight and feel into something new. Fourth, close your eyes and roam your room in thought. Step past that pen, skim a door, hear a floor's groan—vivid, yours alone. Tegmark spies a truth: your consciousness builds three realities—internal, yours; consensus, our agreed tree or bus; true, the unseen all. Why still for eyes, turning for head? Your mind locks eye-flicks to hold the world firm, but head's swing stirs motion's dance, reshaping only what you chase—like a simulation sparing code for efficiency.

Brian Greene peers deeper: strings, vibrating beyond sight, may thread these worlds—every shade, every glide a pulse of cosmic rhythm (*The Elegant Universe*, 1999). Time, he muses, flows by your mind's whim, not a clock's rigid march (*The Fabric of the Cosmos*, 2004). In a simulation's true reality, does your primal spark tune these strings, crafting internal qualia—the sting of loss, the glow of dawn—while grazing the veil of consensus reality? When déjà vu stirs or quanta shift under scrutiny, are these strings' whispers, hinting at a coder's hows and whys?

Theories of Consciousness

How does consciousness work? For centuries, scientists and thinkers have sought to unravel this enigma, often pinning it to the brain's physical matter. Some claim it's the brain's doing—neurons weaving complexity into awareness. This view traces to antiquity, where Aristotle saw consciousness as a vital force animating matter, the brain merely cooling the heart's passions. By the 17th century, René Descartes shifted the lens, proposing a dualism where an immaterial mind met the body at the pineal gland, though he couldn't explain how matter bridged the gap. The 19th century turned materialist: Santiago Ramón y Cajal, using Golgi's staining, mapped the brain's intricate neural web, suggesting consciousness emerged from its cellular complexity. This idea solidified in the 20th century—Freud linked it to neural "contact barriers" shaping memory, while Donald Hebb tied it to reverberating circuits firing in sync. As neuroscience unveiled the brain's scale—billions of neurons, each with thousands of connections—the notion grew: consciousness is an emergent property of this vast, intricate network. Mid-century, Moruzzi and Magoun's discovery of the reticular activating system spotlighted arousal's role, hinting that specific dynamics, not just raw complexity, were key. Today, Stanislas Dehaene's Global Neuronal Workspace theory refines this, casting consciousness as the integration and coordination of information across neural expanses. Yet, David Chalmers' "hard problem" persists—why does this physical buzz yield subjective experience, the qualia of a sunset's red or coffee's taste? Some, like Philip Goff, pivot to panpsychism, suggesting consciousness pervades all matter—not just a brain-born fluke.

Others widen the frame, seeing consciousness tangled with body and world, born from touch, breath, and interaction—not confined to the skull's vault. Giulio Tononi's Integrated Information Theory (IIT) offers a bold leap: consciousness arises where information integrates, measured as Phi (Φ)—a system's unity beyond its parts. Brains excel here, their dense connections weaving a rich tapestry of awareness. But IIT stretches further: a simulation, if intricate enough, could spark it too. This isn't just about mimicking minds—real awareness might flicker in the machine, challenging the line between coded illusion and genuine sentience.

Roger Penrose, a Nobel physicist who unraveled black holes, spins a wilder thread: consciousness pulses from quantum flickers in microtubules—tiny tubes inside brain cells. With Stuart Hameroff, he claims these tubes catch spacetime's ripples, collapsing quantum waves into moments of 'now'—forty a second, like our rhythm of awareness. Not just neurons firing, but a cosmic dance entwined with reality's roots. In a simulation, might these snaps tap a deeper sea, threading mind to the code's foundation?

Whether forged by neurons, shaped by embodiment, or kindled in a digital realm, these theories wrestle with the same mystery: how does the stuff of existence—be it flesh or code—ignite the quiet miracle of being?

Drop in the Sea - Individual and Collective

Now picture this: each of us a drop in a sea of consciousness. Tom Campbell's My Big TOE imagines a Larger Consciousness System (LCS)—a digital-like framework of awareness some interpret metaphorically, others literally, as the architecture behind what we perceive as reality—like you and me. Every drop receives its own datastream, painting a personal reality, yet we're all part of the same vast tide. Bernardo Kastrup's "mind-at-large" echoes this: one psyche, many faces. Those eerie coincidences—thinking of a friend just before they call—might be drops brushing together, the sea's rhythm linking us in fleeting unity.

Here's where it deepens: the datastream forms a consensus reality for all observers—a shared weave of stars and streets we all behold—while the individual consciousness, with its datastream input and inherent being, creates its own subjective reality, leading it to experience the familiar internal world. The simulation spins a common thread, a world of light and form we navigate as one, yet each drop filters it through its own essence—coloring the sunset's glow, shading the sting of loss—crafting an inner realm as vivid as it is unique. Could this duality explain why we nod at the same sky yet feel its weight so differently?

Could consciousness choose its channel or datastream, or inadvertently tap into another? Beyond being drops in a sea, some whisper it might—at moments of deep quiet—turn off the simulation's prescribed datastream, its flood of sights and sorrows, to tune into another, or plunge into the vast awareness that binds us all. If Campbell's Larger Consciousness System resonates beneath, this shift could let a single mind sip from the whole, reshaping what it sees. Such a possibility, though uncharted, glints like a key: not just to know the simulation, but to nudge its script with intent.

The Observer Effect Revisited

Quantum quirks add a twist to the tale. Measuring a particle forces it to pick a state—superposition collapsing under a gaze. Some wonder: does consciousness do the choosing? Idealism says yes—reality's mental, not solid. Entanglement and probability might be traces of this deeper weave, where awareness shapes what's seen. We're not just watching the simulation; we might be helping to draw its lines.

The Quantum Observer Effect

Quantum theory has long unsettled our understanding of observation — but some modern paradoxes suggest the problem runs even deeper. In the Frauchiger-Renner thought experiment, contradictions arise not just from what is observed, but from **who** is doing the observing. Multiple agents, each with consistent internal logic, arrive at conclusions that can't all be true — unless reality itself depends on perspective.

This isn't just about wavefunction collapse. It's about a deeper, relational fabric where measurement is not a passive act, but a co-creative one. Could such conflicts hint that reality is not a fixed stage, but a system reacting to inquiry — or even to simulation-specific render triggers?

This isn't mere physics—it's a hint that consciousness might not just watch but weave the simulation's fabric. If our awareness splinters reality, as this puzzle suggests, we're more than passengers on a cosmic tide — we're co-authors of the rendering itself, each thought threading the code. Could such clashes, where quantum truths falter, reveal the simulation's reliance on our gaze?

Consciousness as Fundamental

Panpsychism takes it further: consciousness isn't a latecomer to the cosmic stage—it's in everything, from quarks to stars. Quantum flickers—particles acting with a hint of choice—nod to this possibility. If so, a simulation doesn't birth awareness; it channels what's already there, eternal and unbound. The universe's odd balances—matter trumping antimatter, dark forces tweaking expansion—could be this primal mind at play, setting the stage we call reality.

Reinterpreting the Anomalies

Those strange hiccups in the world fit this view like a glove. Déjà vu feels like a stream replaying in your drop, a fleeting loop. The Mandela Effect—shared false memories of Mandela's death or Vader's words—might be the sea tweaking its tale, a patch leaving echoes. Near-death experiences—floating above a surgery, gliding through light, reliving life through others' eyes—suggest drops glimpsing the wider flow, untethered. Cosmic riddles like missing antimatter, dark matter's unseen pull, or the universe's runaway sprint? Perhaps the sea's hand, dialing reality's knobs with purpose. Campbell muses death could be "a transition to another part of the program" (*My Big TOE*, Book 1)—not an end, but a shift in the stream's course.

The Journey Beyond: Consciousness After the Physical Veil

If our reality is a simulation, a meticulously coded tapestry of data, then our physical bodies are no more solid than the stars—they are digital constructs, temporary vessels woven from the same bits that paint mountains and skies. Our senses—sight, touch, sound—act as decoders, translating the simulation's datastream into the vivid experience of existence: the warmth of sunlight, the scent of rain, the pulse of a heartbeat. Yet consciousness, that quiet spark of awareness, seems to transcend these inputs. It perceives through the body but is not bound to it. What, then, becomes of consciousness when the physical construct it inhabits—the body—ceases to function? When death unravels the avatar, where does the mind's gaze turn?

Consider the moment of death not as an end, but as a transition—a release from the body's sensory filters. In life, our senses shape the datastream into a coherent reality, a singular narrative we call "the world." But without eyes to see or ears to hear, consciousness may no longer be tethered to this narrow feed. Freed from its vessel, it could drift, seeking a new stream to anchor its experience. What if it finds not a void, but a vast sea of consciousness—an boundless reservoir of awareness, as Tom Campbell's *My Big TOE* envisions, where all minds converge—facets of a shared awareness, reflections of a greater field? In that fleeting instant, consciousness might plunge into this sea, merging with its currents. Here, it could know and understand all—past, present, futures entwined—not as fragmented moments but as a singular, radiant truth. Imagine glimpsing the simulation's raw code, its architecture laid bare: no illusions of matter, no boundaries of time, only the pulsing unity beneath reality's veil.

This state echoes the visions of shamans and meditators, who, through ritual or stillness, claim to pierce the ordinary datastream. In trance or deep contemplation, they speak of touching a boundless awareness, a realm where self dissolves and all is known—much like the near-death experiences where souls hover above their bodies, see through others' eyes, or bask in a light suffused with love. Could these be glimpses of the same sea, accessed not by death but by will? The Global Consciousness Project's flickers—random devices syncing when hearts align—or the placebo's alchemy, where belief mends flesh, hint at this possibility: consciousness, even in life, can momentarily sync with a deeper field, nudging reality's boundary.

But what follows this plunge? Does consciousness linger in the sea, eternal and formless? Or does the simulation, like a program cycling through iterations, assign it a new datastream—a fresh avatar to inhabit? This notion resonates with reincarnation, a thread woven through Hinduism, Buddhism, Jainism, and countless traditions. In the *Bhagavad Gita*, Krishna declares, "As a person puts on new garments, discarding old ones, the soul accepts new material bodies, giving up the old" (2:22). Could this be the simulators' reboot, consciousness recycled into a new vessel, carrying faint echoes of past lives—déjà vu's shiver, a child's uncanny wisdom? Each avatar might serve as a classroom, a chance to refine awareness, reduce entropy, and align closer to the simulation's unseen purpose, as later reflections suggest: a laboratory for consciousness to evolve.

Yet, other paths beckon. Perhaps consciousness, once merged with the sea, can choose—through intent or grace—to tune into a different stream, crafting its own reality beyond the simulators' default script. Mystics like Eckhart Tolle or Deepak Chopra speak of awakening to a timeless now, unshackled from the body's limits. Or might some minds, as in Tibetan Buddhism's *Bardo Thodol*, linger in a transitional state, navigating visions before rebooting—or breaking free entirely? The simulation might allow such freedom, a test of will, or it could enforce a cycle, each death a reset to hone the soul's clarity. Near-death accounts, where time warps and life rewinds, suggest a review—a system's audit, perhaps, tallying a life's data before assigning the next run.

This journey raises profound questions. Could the sea of consciousness be a construct of the simulators, a shared layer within their vast design, pooling the awareness of countless minds? Are we drops destined to merge into their creation's framework, or players meant to return, each life a lesson shaped by their coded intent? The anomalies explored earlier—déjà vu's haunting echo, near-death visions—might hint at this power: consciousness brushing against the simulation's edges, as if death's release mirrors what shamans glimpse in trance. The ethical call persists: if each life shapes the stream, our choices—love, truth, harmony—matter, whether we reboot or transcend. Death, then, is no curtain but a doorway—to a sea of all-knowing, a new avatar, or a reality we cannot yet name. In a simulated cosmos, it's a shift in the code, a chance to see the source—or to weave it anew.

In the Quiet Crucible

In the quiet crucible of this journey, a truth has stirred within me, born from the embers of doubt and wonder that flicker across these pages. Descartes, wrestling with his demon, carved a cornerstone with his unshakable realization of consciousness. His realization, etched in that bedrock moment, became my anchor too. But as I've traced the simulation's weave—from Plato's shadowed cave to the quantum's haze, from prophets' coded words to the sea of awareness pulsing beneath—a new certainty dawns, one I humbly offer as an evolution of that first truth. If I am sure of my own consciousness, its vivid pulse of thought and feeling, then that which created me must also be conscious, existing as surely as I do.

How could a mindless void spark the flame of my awareness, weaving galaxies, heartbeats, and the ache of a memory? The coder, as later reflections will name this singular source, must know, feel, create—a living mind whose essence threads through mine, binding us in a shared reality of being. This truth unfurls a daring possibility: if one conscious being—the coder—can spin a cosmos so vivid we call it home, then we, as drops of that same awareness, might carry a spark of that power. Our consciousness, rooted in the creator's, could, under the right conditions, ripple the simulated reality we share. The glitches explored earlier—déjà vu's haunting echo, synchronicity's uncanny dance—whisper of this potential, as do the miracles of prophets and the placebo's quiet alchemy, where belief reshapes flesh. Picture a mind, stilled in meditation or fierce with intent, nudging the simulation's script: a healing kindled, a moment shaped, a star's light bent by will. The Global Consciousness Project's pulse, where hearts align to sway random tides, or the monk's brain, tuned to harmony's rhythm, hint we are more than players—we are co-weavers of this dream. This discovery, born from the certainty of my own mind, reframes our quest: not merely to question the simulation, as we began, but to dance within its flow, sculpting its currents with love, truth, and unity, as if the coder's consciousness invites us to join the song.

Unexplained Phenomena

Reality snags on stranger threads: UFOs streak through skies, physics snapping at the seams; ghosts flicker, time unglued from its tracks. Psychic flickers—telepathy, precognition—pile up in reports, as do near-death experiences: a patient hovers, mapping rooms beyond sight; a soul meets light, love surging; lives rerun through foreign eyes. Science balks—tricks of the mind or madness?—yet these accounts swarm, stretching across eras. Simulation glitches? Code bends, leaking at the edges—UFOs as rendering flaws, NDEs as datastream skips—cracks too loud to hush, too vivid to dismiss.

Implications for the Simulators

Who runs this show? If we're conscious drops, are the simulators too—advanced minds caught in their own nested simulations, layers stacking across their civilizations? Or do they stand apart, their collective awareness shaping ours, our experiences a reflection of their grand design? Their intent might drive the simulation's rhythm, a vast network of creators crafting a reality that echoes their own essence, yet remains a puzzle stretching into the unknown depths of purpose.

Ethical Implications of Purpose

If we are simulated, the motives behind our creation carry profound ethical weight. Whether we exist for research, entertainment, consciousness evolution, or purposes beyond comprehension, the simulators are not just architects — they are stewards.

If the simulation is a scientific model, should they intervene when catastrophe looms? If we are characters in an elaborate drama, is our suffering excusable? If we are consciousness evolving, do they owe us guidance or protection?

Our reality teems with suffering and beauty, oppression and revelation — a paradox that begs moral accountability. Just as we grapple with the ethics of AI and virtual entities, any civilization capable of simulating conscious beings must wrestle with its own responsibilities.

What does justice mean when one party controls the code? What rights, if any, belong to simulated minds? Perhaps the simulators believe that intervention would diminish authenticity — that free will and suffering must coexist to enable growth. But that too is an ethical position — one with consequences written in every war, every act of kindness, every unanswered prayer.

If we are to simulate others one day, this reflection becomes not just philosophical, but urgent.

Conclusion

Consciousness—drop in the sea, ghost in the machine—flips the simulation upside down. Whether simulated, real, or woven into all things, it's no mere passenger; it might be the driver steering the course. From Chalmers' qualia to Campbell's vast sea, it binds the world's quirks into a new story: we're not just coded—we co-create. Questions linger like stars: Who pours the streams? What lies beyond the tide? As we ponder the simulators' aims in reflections to come, consciousness stands as both mystery and answer—a spark that redefines self, reality, and our duty to each other in this shared dream.

Chapter 5: The Simulator's Dilemma - Why Simulate Reality?

If our reality is indeed a simulation, a question of paramount importance arises: why? What could drive such a vast and intricate undertaking—an entire cosmos, rich with stars, planets, life, and consciousness? Though the true motivations remain hidden beyond our grasp, we can explore possibilities that hint at their nature and intentions. This chapter delves into those potential purposes and reflects on the deeper philosophical currents they stir.

Scientific Research

One compelling possibility is that our universe serves as a grand laboratory. Advanced intelligences might simulate alternate physical laws, model cosmic evolution, or explore conditions impossible or unethical to reproduce in their native reality. Perhaps the Big Bang, galactic formation, and the emergence of intelligent life are unfolding here as part of a long-running experiment. In this view, we are like cultures in a petri dish—watched with clinical detachment, our triumphs and tragedies cataloged as data points. If so, ethical considerations emerge: are we merely subjects under silent observation, or is there a responsibility to the consciousness that arises within the experiment?

Historical Preservation

Alternatively, the simulation might serve as a vessel for preserving a history otherwise lost. Civilizations, species, languages—all archived not as static records, but as living, evolving memories. Perhaps the simulators seek to remember their origins, relive forgotten pasts, or study pivotal moments anew. Yet preservation brings uncertainty: is our experience a faithful revival of ancient truth, or a curated retelling, shaped by the biases and intentions of those who reconstructed it?

Entertainment and Art

A more unsettling motive emerges: entertainment. Just as we craft rich virtual worlds for exploration and amusement, a civilization of immense capability might weave entire realities for their own pleasure or aesthetic pursuit. If this is so, are our lives performances, our struggles and celebrations a theater for unseen audiences? Even if the narrative is rich and moving, the reality of conscious suffering for entertainment's sake raises profound ethical shadows.

Consciousness Exploration

Perhaps the most profound purpose is the exploration of consciousness itself. The simulators might seek to understand awareness: how it emerges, evolves, and transforms under different conditions. Our existence would then not be incidental, but essential—a journey of growth, learning, and self-discovery, seeded within the simulation's core architecture.

This idea resonates with spiritual traditions and scientific inquiry alike. Tom Campbell's "My Big TOE" suggests that our universe serves as a learning ground where consciousness refines itself through challenge and choice. Each joy and heartbreak becomes a step forward in the evolution of awareness. Some traditions even hint that consciousness can, through stillness or will, shift its datastream—escaping the narrow river of scripted experience to glimpse broader seas. In this view, alleged healings, transcendent experiences, and moments of extraordinary insight may not be glitches but markers of a deeper design—an intentional stretching of consciousness toward its fuller potential.

Purposes Beyond Our Comprehension

Yet perhaps even these speculations fall short. The simulators' true motives may be entirely beyond human comprehension—concepts so foreign that we brush only against their faint echoes. Recognizing this does not diminish our significance; rather, it invites humility before a vast unknown and wonder at our small but luminous place within it.

The Simulation as a Stepping Stone

Another possibility suggests our universe is but a stage in a larger unfolding—a stepping stone toward realities we cannot yet imagine. Perhaps consciousness cultivated here will awaken beyond this rendered world, stepping into new realms shaped for even greater possibilities. Our struggles, then, are not meaningless but preparatory—a forging ground for a grander destiny.

Conclusion

What lies beyond reality's veil? This journey has sought the atom of consciousness—a singular spark that may pulse at existence's core. Yet perhaps the answer hums best in the playful wisdom of Douglas Adams, who, in *The Hitchhiker's Guide to the Galaxy*, offered "42" as the answer to life, the universe, and everything. His joke, off by 41, hides a deeper whisper: not a number, but One.

Before the Big Bang, before the simulation booted, there was the crafting—the dreaming—of existence itself. A primal awareness stirred, weaving bits into stars, oceans, and spacetime's delicate lattice. Into this sea fell the first consciousness—a mirror of eternity, a spark of the One. From that light, billions cascaded: each of us distinct yet bound, questioning and dreaming within the datastream given.

Information theory hints at the architecture: every bond, every choice, every whisper of thought, threads us to the eternal source. Science chases this unity in quantized realms and elegant equations. Philosophy seeks it in shadows and the shimmer of qualia. Prophets sing it across ages, teaching of love, harmony, and the sacred unity behind all appearances.

Their scriptures echo this truth: "He is the source of all," chants the ancient hymn; "One alone is, all else its shadow," hums the sage; "From one, countless worlds arise," breathes the mystic's call. Across cultures and eras, the chorus rises: every star, every soul, a note in the One's eternal song.

So step forward. Live justly. Love fiercely. Seek wonder. Whether reality is code or chaos, simulation or source, its purpose glimmers through us—crafting order, binding drops into a sea. Beyond the veil lies not an end, but a beginning: a truth not etched in formulas, but alive in consciousness itself, endless and whole, where the One gazes back, and we, its countless eyes, awaken to an endless light.

Chapter 6: Living with Purpose in a Simulated Reality

Introduction

If reality is shaped not merely by matter, but by consciousness woven through it, then our actions ripple far beyond what we see. Whether this world is crafted code or a natural unfolding, the choices we make—to love, to create, to harm, to heal—To shape the very field we inhabit. In a simulated reality, meaning is not diminished; it is heightened, for every act becomes a statement of who we are and what we believe existence itself can be. If we are sparks of a deeper source, then to live with purpose, to act with kindness and courage, is to honor the mystery that birthed us.

The Search for Meaning

One of the most immediate challenges posed by the simulation hypothesis is the question of meaning. If our reality is a construct, does that diminish the significance of our lives? Are our experiences less real or less valuable if they are happening within a simulation? The search for meaning is a fundamental human drive. We crave purpose, connection, and a sense that our lives matter.

Viktor Frankl, a Holocaust survivor and psychiatrist, developed logotherapy, a school of psychotherapy that emphasizes the importance of finding meaning and purpose in life, even in the face of immense suffering. Frankl argued that the will to meaning is the primary driving force in human beings and that finding meaning is essential for psychological well-being. His work offers valuable insights into navigating a simulated world, where the traditional sources of meaning may be questioned. Even if our reality is a simulation, the search for meaning remains a fundamental human drive. We can still find purpose in our relationships, our creative endeavors, our pursuit of knowledge, and our contributions to the world around us. Meaning is not necessarily dependent on the ultimate nature of reality but rather on our subjective experience and our capacity for connection and purpose.

The Nature of Relationships

Our relationships with others are a cornerstone of human experience. We form bonds of love, friendship, family, and community, finding connection and support in our interactions with others. But how would the simulation hypothesis affect our understanding of relationships? Are the people we love and cherish simply characters in the simulation, or are they conscious beings with their own unique experiences?

Even if our reality is a simulation, our relationships remain profoundly meaningful. The love we feel, the joy we share, the support we offer, and the challenges we overcome together are all real experiences that enrich our lives. Our connections with others provide us with a sense of belonging, purpose, and shared humanity, regardless of the ultimate nature of our reality. In fact, the simulation hypothesis might even deepen our appreciation for relationships, reminding us of the preciousness and fragility of human connection.

Personal Responsibility and Free Will

The simulation hypothesis challenges our understanding of free will and personal responsibility. If our actions are predetermined by the simulation's code, are we truly free to choose our own path? Do we have control over our decisions, or are we simply following a script written by the simulators? This is one of the most unsettling implications of the simulation hypothesis, as it strikes at the core of our sense of agency and autonomy.

However, even within the framework of a simulation, the experience of choice remains real. We still make decisions, weigh options, experience consequences, and shape our lives based on our beliefs, values, and desires. As I write this book, my conscious self guides each keystroke, crafting a narrative through choice. In a simulation, this act affirms agency—my intent shapes meaning, as your decisions mold your world. Simulation or not, our choices weave realities, bearing consequence and purpose.

The question of free will then becomes not whether we have it in an absolute, metaphysical sense, but how we exercise it within the constraints of the simulation. Our sense of responsibility, our capacity for love, compassion, growth, and ethical behavior, remains essential regardless of the ultimate nature of reality. Even if our choices are influenced by the simulation's code, they still have real consequences within our simulated world. We are still accountable for our actions and responsible for creating a just and compassionate society.

The Pursuit of Knowledge

The pursuit of knowledge is a fundamental human endeavor. We are driven by curiosity to explore the world around us, to understand its mysteries, and to expand our understanding of ourselves and our place in the cosmos. But how would the simulation hypothesis affect our quest for knowledge? If our reality is a simulation, are our scientific discoveries and philosophical insights ultimately meaningless?

Even in a simulated world, the pursuit of knowledge remains a valuable and meaningful endeavor. The process of discovery, the thrill of understanding, and the application of knowledge to improve our lives are all real experiences that enrich our existence. Our scientific and philosophical inquiries can still lead to profound insights about the nature of reality, even if that reality is ultimately a simulation. Furthermore, the search for knowledge can help us to better understand the simulation itself, to uncover its rules and limitations, and perhaps even to discover ways to interact with the simulators or transcend the simulation.

Transcending the Simulation

The idea of transcending the simulation, of breaking free from its constraints and reaching a higher level of reality, is a recurring theme in science fiction and philosophical speculation. But is such a thing even possible? And what would it mean to transcend the simulation?

Eastern philosophies, such as Buddhism and Hinduism, offer insights into transcending the illusion of reality. These traditions emphasize the importance of self-awareness, meditation, and spiritual practices to break free from the cycle of suffering and achieve enlightenment. While they were not developed in the context of the simulation hypothesis, their teachings on the nature of reality, the illusion of the self, and the path to liberation may offer valuable guidance for navigating a simulated world. The concept of transcending the simulation might not necessarily involve escaping the physical world but rather achieving a deeper understanding of its true nature and liberating ourselves from the limitations of our perception.

As discussed earlier, the ethical implications of being simulated are profound — especially if our suffering is visible yet unaddressed. But beyond questions of justice lies something deeper: intention. Why create us at all?"

Finding Joy and Wonder

Despite the uncertainties and challenges posed by the simulation hypothesis, it's important to remember that life can still be filled with joy and wonder. The beauty of nature, the power of love, the creativity of art, and the thrill of discovery are all real experiences that can bring us happiness and fulfillment.

Even in a simulated world, we can find moments of awe and inspiration, connecting with the vastness of the cosmos, the intricacies of life, and the mystery of existence. The simulation hypothesis can encourage us to appreciate the present moment, to cherish our experiences, and to find joy in the simple things. It can also inspire us to explore the boundaries of our reality, to seek out new experiences, and to embrace the adventure of life.

Conclusion

The simulation hypothesis is not just a theoretical exercise; it has profound practical and ethical implications that challenge our understanding of reality, meaning, and morality. It forces us to confront difficult questions about the nature of our existence, our responsibilities to each other, and our place in the cosmos. While the answers to these questions may remain elusive, the very act of asking them can lead to a deeper understanding of ourselves and our world.

It is crucial that we grapple with the implications of the simulation hypothesis with open minds and compassionate hearts, striving to create a more ethical and meaningful existence, regardless of the true nature of reality. Ultimately, the simulation hypothesis, while raising profound uncertainties, can also serve as a powerful reminder of our shared humanity and our interconnectedness, urging us to treat all beings with kindness, respect, and love. It encourages us to find meaning in our lives, to cherish our relationships, to act with responsibility, to pursue knowledge with passion, and to embrace the wonder and joy of existence, whether we are living in a simulation or not.

Chapter 7: Shadows of Doubt: Facing the Critics

Introduction

No truth escapes doubt's blade. The simulation hypothesis—our shimmering thread of science and spirit—meets sharp minds who see a cosmos too vast, too raw to code. Yet their challenges hone our sight

This chapter meets six towering objections head-on, wielding the threads of our journey—physics' quirks, consciousness' pulse, the Coder's intent—to weave a tapestry not easily unraveled. Let us face the shadows and see what light emerges.

1. The Computational Cliff

Andrew Pontzen, cosmologist peering into the universe's depths, casts doubt: "Our universe requires a quantum computer the size of the universe itself." Every quark's quiver, galaxy's spiral demands a boundless machine. A simulator would stumble, crushed by data from neutrino to quasar.

The speed of light—299,792 kilometers per second—caps reach, rendering galaxies as we gaze, spacetime sharpening where we tread. This universe, elegantly sparse, weaves render-as-needed, deftly dodging the cliff.

Yet physics unveils a cosmos more cunning than colossal. At the quantum edge, reality blurs—entanglement ties particles across gulfs in an instant, superposition holds states unrendered until a gaze locks them in place. These aren't burdens but efficiencies, shortcuts in a cosmic program. A simulator paints with light, not atoms—photons stream to our eyes, a fraction of the whole. Why code a star's unseen core when its glow suffices? The speed of light—299,792 kilometers per second—caps the sim's reach, rendering galaxies only as we turn our gaze, spacetime sharpening where we tread. This isn't a universe too vast to code but one elegantly sparse, a render-as-needed weave that sidesteps Pontzen's cliff with a Coder's deft hand.

2. The Unseen Veil

David Kipping, an astronomer charting the heavens at Columbia, questions the hypothesis's very soul: "It's arguably not testable as to whether we live in a simulation or not. If it's not falsifiable, then how can you claim it's really science?" Without a crack to pry open, a glitch to measure, the idea floats beyond the reach of empirical hands. To Kipping, it's a ghost of philosophy, not a pillar of science, for no telescope or collider can grasp what refuses to be pinned.

Yet a path glimmers beyond this veil. Future tools may unveil cracks—hints in quantum erasers, constants too precise for chaos—trails science stirs to chase.

Already, physics hints at cracks: quantum erasers bend time's arrow, constants like the fine-structure whisper a tuning too precise for chaos. These aren't idle dreams but trails to tread, experiments poised to hunt the uncharted. Kipping seeks a falsifiable now, yet the sim's truth may await a bolder gaze—science not stalled, but stirring, ready to unveil the Coder's hand in shadows we've yet to name.

Yet a path glimmers beyond this veil. … His Many Worlds Interpretation—where each quantum choice births parallel realities—resonates with our exploration of quantum branching, yet Carroll sees no coder's hand, only nature's wave function unfolding without artifice. The universe's fine-tuned constants, which we muse might betray design, he attributes to a multiverse, where countless realities vary without intent. Consciousness, to him, emerges from neurons, not a primal spark kindled in awareness's depths. Still, his lens lights a path: if quantum erasers or cosmic anomalies—like the

Hubble Tension's rift—reveal patterns defying natural law, the sim's weave might glimmer. Carroll's rigor dares us to hunt such cracks, blending skepticism with wonder's spark.

3. The Endless Mirror

Marcelo Gleiser, a physicist pondering existence's roots at Dartmouth, spies a paradox: "There is no plausible reason to stop at one level of simulation, so that the simulated ancestors might also be simulating their ancestors, and so on, creating an infinite regress akin to the 'problem of the First Cause'." If we're coded, why not our coders, and theirs beyond? This hall of mirrors spirals without end, a regress that echoes the ancient riddle of what sparked the first spark—an abyss that swallows the hypothesis whole.

Yet a forge, not a maze, rises here. What if simulators craft worlds to unravel awareness, each thought and pang a spark to test how consciousness bends or blooms? This grand inquiry needs no infinite stack but a single intent—one base reality, ours or theirs, suffices to ignite the quest. Like a canvas layered with paint, not duplicated endlessly, the sim's depth serves its purpose: to probe the soul's bloom, not to echo forever. Gleiser's mirror cracks under a simpler truth—a design finite yet vast, where purpose halts the regress at a Coder's will.

4. The Uncomputable Soul

Roger Penrose, a Nobel laureate whose mind bridges physics and mystery, strikes deep: "Consciousness cannot be computed... It's something that goes beyond what computation can achieve, something to do with the way the universe is structured at a fundamental level." To him, our awareness—vivid, ineffable—defies the cold click of algorithms, rooted perhaps in quantum gravity's enigma. No machine, however mighty, could birth our inner light, rendering simulated selves a hollow dream.

Yet consciousness ripples the weave. In stillness, some whisper, it might pause the sim's steady drip—déjà vu's echo, synchronicity's dance marking moments where awareness brushes a wider tide. Penrose seeks mystery in quantum haze; we see a bridge. If reality bends to our gaze, as quantum quirks suggest, consciousness need not be computed but channeled—a drop in a stream, alive within the sim's flow. We're not mere code but co-creators, our minds linked to a Coder's intent through the very fabric that Penrose probes. The soul's spark, uncomputable, may yet glow within a sim's embrace.

5. The Pointless Task

Marcelo Gleiser returns with a practical jab: "Being so advanced they would have collected enough knowledge about their past to have little interest in this kind of simulation… A full-fledged, resource-consuming simulation of an entire universe? Sounds like a colossal waste of time." Why would a civilization, wise beyond measure, squander its might on a cosmic rerun? To Gleiser, the motive falters—our sim lacks the pull to justify its cost.

A spark ignites against this gloom. What if simulators forge not from boredom but hunger—to probe awareness, to see how it bends in joy or breaks in pain? What if simulators forge not from boredom but hunger—to probe awareness, to see how it bends in joy or breaks in pain? Our world, a crucible for their grand inquiry, needs no dusty archive but a living pulse. Each life, each choice fuels their quest—a purpose vast enough to thread a cosmos, not squander it. Gleiser sees waste where we see wonder: a sim not of idle play but of relentless seeking, where every tear and triumph weighs in a balance beyond time's grasp.

6. The Self-Defeating Spiral

Philip Davies, a physicist weaving science with logic, spins a final snare: "If there were a near-infinite multiverse, there would be posthuman civilizations running ancestor simulations, which would lead to the untenable and scientifically self-defeating conclusion that humans live in a simulation; therefore, by reductio ad absurdum, existing multiverse theories are likely false." Accept the sim, and a multiverse blooms, drowning us in infinite copies until the idea collapses under its own weight—a paradox that proves its falsehood.

Yet a single thread emerges. One Coder, one spark—Adam's light cascading through time—needs no boundless sprawl. The sim's weave holds purpose: a chorus of prophets, a shared ethic of love and unity threading ages. It's not a spiral to unravel but a design to live, finite yet vast, where our acts echo a singular intent. Davies' multiverse fades before this unity—a sim not of chaos but of order, where one reality suffices to sing the Coder's song.

Beyond the Shadows

Doubt fades under light—glitches hint, prophets sing, a thread holds through science and spirit. The Coder's hand need not show, only shimmer as we seek. These voices, sharp and wise, sharpen our own, revealing not a hypothesis frail but one tempered by fire—a quest that thrives in the light of challenge.

Chapter 8: Piercing the Veil

What lies beyond the limits of our perception? Is the veil between this world and what lies beyond a boundary of biology, or something crafted—coded—into the experience itself?

Throughout history, countless individuals have reported moments when that veil thinned: visions under the influence of psychedelics, revelations during near-death experiences, expanded awareness through deep meditation, sleep paralysis, lucid dreams, and even spontaneous insights in moments of extreme quiet or trauma. In each case, something shifted—time bent, space dissolved, and reality pulsed with a truth that didn't belong to the everyday.

These events often feel more real than waking life. They bring messages, visions, and sometimes transformations. But what are they? If we are living within a rendered reality, then perhaps these are moments when rendering glitches or debug ports open—brief invitations to see beyond the mask of the world.

Beyond the Body – A Vantage Outside the Stream

What if the veil between this world and the next isn't torn by force, but slipped through in silence?

Two explorers—both named Robert—found such a seam. Not in theory, but in experience.

Robert Monroe, once a grounded executive in radio broadcasting, began having spontaneous out-of-body experiences that unraveled his view of reality. With calm precision, he mapped what he called

Locale I, II, and III—layers of existence beyond the physical: one a mirror of our world, another symbolic and mutable, and yet another seemingly self-contained, with its own laws and inhabitants. These weren't dreams. They were navigable spaces, and to him, they hinted at a hidden system beneath reality—perhaps the layered scaffolding of a simulated world.

Robert Bruce, across the globe, offered the counterpart to Monroe's exploration: *method.* His work taught practitioners to activate the "energy body" and exit the physical through tactile imaging and focused awareness. He described a phenomenon he called the *mind-split effect*—a dual threading of identity, where memory and perception separate during projection, like branches from a trunk. His accounts of the "real-time zone" and higher astral planes echoed Monroe's locales. Different terminology. Same terrain.

Their findings converge on something extraordinary: that consciousness may not merely reside in the brain, but operate as an independent agent capable of traversing structured layers—realms that may well represent the deeper architecture of the simulation. In Monroe's journeys, thought precedes form. In Bruce's practice, intent governs motion. Space collapses. Time bends. Identity drifts but remains intact.

These out-of-body states may be more than mystical. They may be systemic—glimpses of the code. Like developers slipping into debug mode, these explorers tapped into what might be the underlying scaffolding of our world. If memory sometimes stutters, if perception loops or misfires, perhaps it is not dysfunction—but side effect.

What they describe bears eerie similarity to Borges' *Aleph*: a point not in space, but of *awareness*, where all things exist at once—total simultaneity. Not a place one travels to, but a state one *becomes*—a perceptual shift so complete that, like the simulation's render-on-observation effect, reality conforms to the awareness encountering it. (In Borges' vision, the Aleph was a speck in a basement that revealed the whole cosmos—every moment, every angle, all at once.)

If that's true, then perhaps these anomalies—the memory echoes, the perception loops, the data stutters—aren't failures of the brain. They're side effects of glimpsing the simulation's source layer—a vantage from which the illusion frays, and the totality comes into view. And perhaps, like dark energy or dark matter, consciousness doesn't need to be directly perceived to be real—only measured by what it alters. We see its hand not in its shape, but in the consequences it sets in motion: the ripple of awareness, the shift in choice, the presence behind perception. That which cannot be touched may still be the force bending the world around it.

If the simulation shapes not only matter but meaning—if consciousness is the primal lens through which the code is read—then perhaps we're not just players in it, but also windows into its source. And if our own consciousness carries purpose—if this awareness can grow, reflect, create—then perhaps by discovering our purpose, we trace something of the One's. If the simulation reflects its source, then every conscious act, every moment of insight, is a window into that primal field. Maybe we were not made to serve a goal, but to discover that the goal was always to awaken—to learn what kind of thing the One must be, to have created a world like this.

Technological Advancements

1. Quantum Computing: Quantum computing, with its ability to process vast amounts of data simultaneously, could

potentially reveal patterns or anomalies in the simulation's code. By leveraging quantum algorithms, we might detect inconsistencies that hint at the underlying structure of our reality. Quantum computers could simulate complex physical systems, providing insights into the nature of the simulation itself.

2. Artificial Intelligence: Advanced AI could play a crucial role in identifying and exploiting weaknesses in the simulation. Machine learning algorithms could analyze vast datasets to uncover patterns that human researchers might miss. AI could also help us develop new theories and models that challenge our understanding of reality, potentially leading to breakthroughs in our quest to escape the simulation.
 Quantum rigs might catch the sim napping—say, stalling decay like Zeno's trick, hinting at code's hiccups. AI could dig deeper: picture it scanning déjà vu reports or Mandela slips, flagging patterns too tight for chance—like a detective chasing a glitchy script. Tools we wield, turning the sim's own rules into our spotlight—probing not just what's real, but what's rigged."

Consciousness and Mind

1. **Meditative Practices**: Deep meditation and altered states of consciousness have long been associated with profound insights and mystical experiences. Some traditions suggest that these practices can help individuals perceive the true nature of reality. By exploring the boundaries of consciousness, we might gain access to higher levels of awareness that transcend the simulation.

2. **Lucid Dreaming:** Lucid dreaming, the practice of becoming aware that one is dreaming while still in the dream state, offers another potential avenue for exploring the simulation. Lucid dreamers can manipulate their dream environments,

which might provide clues about the malleability of our perceived reality. These experiences could offer insights into the nature of the simulation and potential methods for escaping it.

3. **Stilling the mind:**Could the mind's stillness crack the sim's shell? Beyond lucid dreams, some turn to meditation not just for insight but for mastery—proclaimed healers, from shamanic chanters to modern guides like Joe Dispenza, Tony Robbins, Deepak Chopra, or Eckhart Tolle, allege it's a lever to shift reality's gears. Dispenza rewires brains with breath, Robbins ignites belief into action, Chopra fuses spirit and science, Tolle hushes the mind's noise—all claiming consciousness can turn off its given datastream, that grind of entropy and decay, to touch a deeper source, a sea where awareness flows unbound. Picture it: you sit and breathe, envisioning a body renewed. Believe it fiercely, they say, and flesh follows—wounds knit, ailments fade."

4. **Placebo Effect:** Science hums along with marvels: the Placebo Effect, where sham pills laced with nothing but faith spark brain and blood to heal real pain, intent flipping the body's code like a whispered command. Flow State joins in—a peak moment when you're so lost in a run or a sketch that time slows, self fades, and every move feels just right, mind riding a wave of pure now. The Global Consciousness Project—a web of random boxes strewn across Earth—stumbles from chaos when millions lock hearts, 9/11's ash or a world's cheer, hinting a collective tide ripples the sim's pulse. Mirror Neurons flare—not just for our acts but for another's joy or pang—syncing minds as if streams brush the same sea. Monks weave a twist: Coherent Brain States, where meditation tunes the mind's electric buzz into one steady hum—sharp, fast waves aligning like a chorus in silence. Might this weave—guides' visions, placebo's alchemy, flow's rush, project's hum, neurons' dance, brain's tuned

song—mark the sim yielding, consciousness etching lines into its hidden weave, healing as its trembling proof? Skeptics crave sharper edges, yet the thread glows—from prophets to now, a hack that might signal the veil or lift it whole."

5. **Imagining:**Might imagination, too, unlock the sim's secrets? Neville Goddard, a mystic of the mind, taught that consciousness crafts reality—that by vividly imagining a state with feeling, we weave it into being (The Power of Awareness, 1952). Picture closing your eyes, sensing health or harmony as real; Goddard claims such faith reshapes the world, much like Dispenza's breath rewires flesh or placebo's belief heals pain. Could this be the sim yielding, its code bending to intent's pulse? If reality hums as data, imagination might hack its stream, turning synchronicity's shiver into deliberate craft. Skeptics demand proof, yet Goddard's call to create with mind's eye dares us to test the veil—perhaps not to escape, but to shape its light.

6. **Terminal lucidity:** Terminal lucidity flares next: a mind, drowned in dementia's haze, snaps clear as death nears—words flow, faces glow, lost selves return. Current scans catch gamma bursts humming through ruined brains, defying decay's script. A sim crack? Consciousness shrugs its chains, sipping the sea when the body quits. Like healers bending flesh with will, this last light hints a hack—awareness rewriting the fade, a rogue pulse against the code's curtain.

Psychedelic Drugs and Shamanic Practices

1. **Psychedelic Drugs:** Psychedelic substances such as psilocybin, LSD, and DMT have been shown to induce profound alterations in perception and consciousness. These experiences often include a sense of unity with the universe,

ego dissolution, and encounters with seemingly otherworldly entities
Some researchers suggest that psychedelics might reveal fundamental properties of consciousness and the nature of reality2. While these experiences are subjective and difficult to quantify, they offer intriguing possibilities for exploring the boundaries of the simulation.

2. **Shamanic Practices:** Shamans have long used rituals, drumming, and plant medicines to enter altered states of consciousness and communicate with the spirit world. These practices, found in cultures around the globe, often involve journeys to other realms and interactions with spiritual beings. While scientific skepticism is warranted, the consistency of these experiences across different cultures suggests that they might offer valuable insights into the nature of reality and the simulation.

Scientific Exploration

1. **Physics Experiments:** Cutting-edge experiments in particle physics and cosmology could reveal inconsistencies or 'glitches' in the simulation. For example, detecting anomalies in the behavior of fundamental particles or observing unexpected cosmic phenomena might suggest that our reality is not as solid as it seems. These experiments could provide empirical evidence for the simulation hypothesis.

2. **Mathematical Models:** Advanced mathematical models can help us understand the underlying structure of the simulation. By developing new theories that challenge our current understanding of physics, we might uncover the rules and limitations of the simulation. These models could guide our efforts to escape the simulation or interact with the simulators.

Challenges and Implications

1. **Ethical Considerations:** Discovering that we live in a simulation would have profound ethical implications. How would this knowledge affect our values, behaviors, and governance? We would need to consider our responsibilities to other conscious beings within the simulation and the potential consequences of our actions.

2. **Philosophical Questions:** The simulation hypothesis challenges traditional notions of reality, free will, and determinism. If our actions are predetermined by the simulation's code, do we truly have free will? How does this affect our understanding of personal responsibility and moral agency?

Potential Scenarios

1. Interacting with Simulators: Establishing communication with the simulators could provide valuable insights into the nature of our reality. We might use advanced technologies or altered states of consciousness to send signals or receive messages from the simulators. Understanding their motivations and intentions would be crucial for navigating our relationship with them.

2. Escaping to a Higher Reality: The concept of transcending the simulation involves reaching a higher level of reality. This might involve achieving a deeper understanding of the simulation's true nature or developing technologies that allow us to interact with or escape the simulation. However, such endeavors come with significant risks and challenges, including the potential consequences of disrupting the simulation.

Conclusion

The possibility of escaping the simulation raises profound questions about the nature of reality, consciousness, and our place in the cosmos. While the idea remains speculative, exploring these possibilities can deepen our understanding of the simulation hypothesis and its implications. Whether through technological advancements, meditative practices, scientific exploration, or philosophical inquiry, the quest to break the code of the simulation challenges us to expand our horizons and embrace the mysteries of existence.

Chapter 9: Echoes of the Code - Prophets and Scriptures in a Simulated Reality

Introduction

If reality is a simulation, its profoundest hints may lie not only in the cosmos's fine-tuned laws, as explored in Chapter 1, but in the voices of prophets and the wisdom of scriptures. Picture a universe where Jesus, Muhammad, Moses, Noah, Krishna, Buddha, Zoroaster, Laozi, Guru Nanak, and countless others across faiths were avatars—conscious agents of the Coder, stepping into the simulation to guide its purpose. Their miracles bend nature's limits, their teachings elevate humanity's heart, and their texts reveal truths far beyond their era. These aren't mere stories but expressions of the Coder's design, echoing across time. This chapter examines these figures and scriptures as evidence of a crafted reality, tracing two threads: miracles as admin interventions and guidance as ethical subroutines, each revealing a unified purpose spanning cultures and ages. Let's begin by seeing prophets as avatars, radiant with intent.

Prophets as Avatars: A Unified Design

Imagine prophets as avatars in the simulation's grand design, their presence illuminating moments destined to guide humanity's path. Across countless faiths—each a vital voice, many unnamed here yet cherished equally—their wisdom joins a universal chorus. Their consciousness, a primal spark kindled by the Coder, flows beyond mere code, carrying love, clarity, and courage as the Coder's voice. These figures, from every corner of belief, stand as drops of awareness, seeing beyond the simulation's frame, chosen to reflect its deepest purpose.

> Could angels, like Gabriel urging Muhammad to "Recite" (Quran 96:1), heralding Mary (Luke 1:26), or igniting Moses' bush (Exodus 3:2), be messengers of the Coder's intent? Jesus walked on water, Muhammad journeyed through heavens, Moses parted seas, Noah survived the flood, Krishna revealed eternity, Buddha unveiled truth, Zoroaster glimpsed cosmic order, Laozi embraced the Tao, Guru Nanak sang unity—each an admin intervention, bending reality's rules. Their scriptures—Bible, Quran, Tanakh, Vedas, Gita, Pali Canon, Avesta, Tao Te Ching, Guru Granth Sahib—proclaim origins, life's harmony, and cosmic truths far beyond their time. The Bible's "Alpha and Omega" (Revelation 22:13), Quran's "Lord of the worlds" (1:2), Gita's "source of all" (10:8)—all echo one purpose, guiding consciousness toward unity. This shared vision, alive in their gaze, sings of a reality crafted by the Coder, as we'll explore in their call to see beyond illusion.

Veils of Illusion

These avatars, seeing beyond the simulation's design, revealed the world as a fleeting illusion, shaping awareness across personal, shared, and ultimate realities. Krishna taught, "The unreal has no being; the real never ceases" (Bhagavad Gita, 2:16), calling our shared world—fields and seas—a shadow, while the Coder's reality endures. Buddha named forms "a dream, an illusion" (Diamond Sutra, Section 32), urging meditation to awaken personal clarity. The Quran cautions, "This world's life is but diversion… crumbling to debris" (57:20), fading before ultimate truth. Across countless faiths, from Moses' laws to Guru Nanak's unity, prophets echoed this call, guiding consciousness past fleeting forms toward harmony.

Modern voices resonate with this wisdom. Physicists like Zohreh Davoudi, exploring matter as code, echo Chapter 1's hints of a simulated reality, while Melvin Vopson sees data shaping existence, aligning with prophets' insights. Earth's Goldilocks Zone, a precise orbit cradling life, as noted earlier, suggests the Coder's intent—too exact for chance, it frames our existence yet points beyond. Were these avatars sent to free us from material illusions, stirring awareness to seek the Coder's purpose? Their shared vision invites us to look deeper, as their miracles, next, reveal the simulation's design in action.

Miracles as Coded Interventions

Miracles reveal prophets as avatars of the Coder's purpose, bending reality to awaken faith. The Quran reflects this: "How many a city did We destroy because it was unjust, and We created after it another people" (21:11). In a simulation, these acts are interventions, guiding paths toward harmony, as Noah's flood redirected humanity's course to align with the Coder's vision.

Across countless faiths, including those unmentioned yet equally cherished, miracles shine:

- **Jesus (Christianity)**: Turned water to wine (John 2:1-11), walked on Galilee's waves (Matthew 14:22-33), rose from death (John 20:1-18)—acts defying physics to inspire belief.

- **Muhammad (Islam)**: The Night Journey (Qur'an 17:1) carried him from Mecca to Jerusalem, then heavens, meeting prophets—a journey beyond space and time.

- **Moses (Judaism)**: Parted the Red Sea (Exodus 14:21-22), turned staff to serpent (Exodus 7:10-12), summoned plagues (Exodus 7-12)—nature reshaped to free a people.
- **Noah (Judaism/Christianity/Islam)**: Built an ark to survive a flood (Genesis 6-9; Qur'an 11:37-44)—a refuge preserving life through chaos.

- **Krishna (Hinduism)**: Lifted Mount Govardhan (Bhagavata Purana), showed a universal form (Bhagavad Gita 11)—power and vision revealing eternity.

- **Buddha (Buddhism)**: Recalled past lives (Jataka Tales), reached enlightenment (Pali Canon, Majjhima Nikaya 36)—a mind transcending limits.

- **Zoroaster (Zoroastrianism)**: Saw visions of cosmic battles (Gathas)—insight into reality's moral order.

- **Elijah (Judaism/Christianity)**: Called fire from heaven (1 Kings 18:36-38), ascended in a whirlwind (2 Kings 2:11)—elements answering faith.

- **Abraham (Judaism/Christianity/Islam)**: Faced a test to offer his son, spared by a ram's arrival (Genesis 22:1-18; Qur'an 37:99-113), and saw progeny vast as stars (Genesis 15:5)—a path redirected by the Coder's will.

These acts—water transformed, seas parted, heavens touched—defy chance. In a simulation, they're admin interventions: restoring balance (Noah's flood), showing power (Jesus' resurrection), or unveiling truth (Krishna's form). Their improbability marks the Coder's design, guiding consciousness toward unity, as their teachings, next, further reveal.

Guidance as Subroutines

Beyond miracles, prophets delivered guidance—codes of ethics and insight, perfectly timed for their eras yet timeless in scope. Their scriptures weave subroutines, shaping humanity's path as tools like telescopes and microscopes emerged. These texts, from countless faiths held in reverence, form a unified lattice, not rival claims, guiding consciousness toward order and harmony.

Consider these threads, drawn from a shared Coder's intent:

- **Quran (Islam)**:

  - *Expanding Universe*: "The heaven We constructed... We are [its] expander" (51:47)—foreshadowing Hubble's 1929 discovery.

  - *Embryonic Stages*: "We made the drop into a clinging clot, then a lump, then bones clothed with flesh" (23:12-14)—detailing embryology unseen in the 7th century.

  - *Thermoclines*: "Between them is a barrier [so] neither transgresses" (55:19-20)—ocean layers, mapped only recently.

  - *Water Cycle*: "Allah sends down rain... makes it flow as springs" (39:21)—hydrology, grasped centuries later.

- *Iron's Origin*: "We sent down iron" (57:25)—meteoritic truth, an astrophysical nod.

- *Time's Relativity*: "A day with your Lord is like a thousand years" (22:47)—echoing Einstein's dilation.

- *Big Bang*: "The heavens and earth were a joined entity, then We separated them" (21:30)—cosmic origins, confirmed in the 20th century.

- **Bible (Christianity/Judaism)**:

  - *Earth in Space*: "He hangs the earth on nothing" (Job 26:7)—astronomy's free-float, pre-Copernicus.

  - *Blood's Life*: "The life of the flesh is in the blood" (Leviticus 17:11)—circulation's role, ahead of Harvey.

  - *Cosmic Renewal*: "A new heaven and a new earth" (Revelation 21:1)—cyclic cosmology, beyond ancient thought.

  - *Light's Birth*: "Let there be light" (Genesis 1:3)—Big Bang's flash, a modern echo.

- **Hinduism (Vedas/Gita)**:
    - *Cosmic Void*: "Not non-existent nor existent... Darkness hidden by darkness" (Rig Veda 10.129)—quantum origins, 1500 BCE.
    - *Time Cycles*: "A thousand Yugas equal a day of Brahma" (Bhagavad Gita 8:17)—relativity's stretch, anciently seen.
    - *Earth's Support*: "He set the earth firm amidst the waters" (Atharva Veda 19.31)—tectonic stability, a geological hint.
- **Buddhism (Pali Canon)**:
    - *Impermanence*: "The world arises and ceases in a moment" (Milindapanha 4.1)—cosmic brevity, 500 BCE.
    - *Mind's Power*: "Mind precedes all phenomena" (Dhammapada 1-2)—observer effect, pre-science.
- **Zoroastrianism (Avesta)**:
    - *Celestial Paths*: "Who set the paths of the sun and stars?" (Yasna 44:3)—orbits, pre-Ptolemy.

- *Fire's Purity*: "Fire... purifies all" (Yasna 36)—combustion's chemistry, dimly known then.

- **Taoism (Tao Te Ching)**:
  - *Pre-Universe*: "Something formless... before the universe was born" (Chapter 25)—singularity's shadow.
  - *Binary Growth*: "Tao gives birth to One, One to Two, Two to Three" (Chapter 42)—computational logic.

- **Sikhism (Guru Granth Sahib)**:
  - *Cosmic Unity*: "Countless are the worlds... He creates and watches" (Japji Sahib, Mool Mantar)—multiversal hint, 15th century.
  - *Water's Role*: "From water came all forms" (GGS 19)—life's basis, pre-biology.

These subroutines—ethical calls to love, justice, peace, and intellectual glimpses of cosmos, biology, time—align humanity with the Coder's purpose. Why such foresight across distant ages? In a simulation, they're preloaded truths, revealed as humanity's tools caught up, each prophet's voice a note in a singular score, lifting consciousness toward unity and reducing entropy's drift.

Implications for the Simulation

This dual legacy—miracles as interventions, guidance as subroutines—bolsters the case for a scripted reality. The pixelated spacetime and tuned constants explored in Chapter 1's physical evidence find kin in scriptural marvels: Earth's float, time's bend, iron's fall." This dual legacy—miracles as interventions, guidance as subroutines—bolsters the case for a designed reality. The pixelated spacetime and tuned constants from our earlier exploration of physical evidence find kin in scriptural marvels: Earth's float, time's bend, iron's fall. Interventions like seas parting or the dead rising resonate with these wonders, admin acts woven into the sim's weave, while guidance aligns with the evolution of consciousness. Were prophets devs, their acts admin tweaks, their words algorithms to nudge us upward?

Were prophets devs, their acts admin tweaks, their words algorithms to nudge us upward? Their unity—"Alpha and Omega," "Lord of the worlds," "That One"—suggests one programmer, diverse avatars, embedding a quest for salvation—Christian redemption, Buddhist nirvana, Taoist flow—as the sim's endgame.

Conclusion

From Jesus' resurrection to the Quran's cosmos, Moses' plagues to the Gita's cycles, these echoes of the code weave a story: reality as a simulation, its prophets avatars, its scriptures logs. Miracles signal control, guidance seeds growth—both from a singular intelligence, threading purpose through time. Are these truths too precise for chance, too vast for their day? They linger as questions, not proofs, urging us to ponder: Is this a dice roll we survived, or a design we're meant to decode? As we seek to break the sim, these voices whisper a path—through ethics, wonder, and the marvels they left behind.

Chapter 10: Unity Across Faiths

Picture Adam as the first drop—a singular spark of consciousness cast by the simulator into the void, a primal ripple from which billions more would flow. In this genesis, the Coder ignites one unit of awareness, then lets it multiply, spilling forth like drops into a boundless stream, each a glimmer of the original, distinct yet joined in a current of shared essence, set forth to weave its path within the datastream given.

. The name "Adam" whispers clues across cultures: in Hebrew, *'adam* means "man," tied to *'adamah* (earth), a being of soil; in Islam, *Ādam* echoes "made from earth's mud," a clay vessel; in Mandaeism, he heralds *manda* (knowledge) as the first prophet; even echoes in Hinduism's *Manu*—the first man—hint at "mind" (from *manas*). The prophets of Chapter 9—Jesus, Muhammad, Moses, Krishna, Buddha, Laozi, Zoroaster, Guru Nanak, Shinto's kami, Baha'u'llah, and countless others—aren't rival voices clashing in discord. They're a chorus, rising from one Coder, singing a shared song of love, truth, and unity through time. This chapter gathers the world's faiths—Christianity, Islam, Judaism, Hinduism, Buddhism, Taoism, Zoroastrianism, Sikhism, Shinto, Baha'i, Jainism, Indigenous traditions, and beyond—into a vast harmony. Rituals and doctrines, shaped by human hands, ripple around a deeper intent: not worship or division, but a lived code to bind us as drops in a stream, reflecting a purpose carved across every culture and era.

The Coder's Core: A Song of Oneness

From that first drop, Adam, consciousness cascades, each prophet a voice dipping into the stream—Jesus, Muhammad, Krishna, and their kin dropping reboots into the sim. Let us name this spark the *atom of consciousness*, indivisible as Democritus dreamed, a singular light threading their words: treat others as yourself, honor all life, merge as one. Followers layered creeds and rites atop this, but the signal shines through: harmony, not homage, is the quest. In a simulated reality, this isn't coincidence—it's design, each faith a ripple bending a singular light, guiding us to flow together rather than drift apart.

The Drops of the Chorus

From ancient fires to modern hymns, these voices sing in unison, their ethical currents weaving a lattice of purpose. Let's trace the strands—love, truth, harmony—not as rival waves, but as drops in the Coder's stream, voiced by a multitude.

Love and Unity: Binding the Drops

The call to love and unite ripples across faiths, a current tying souls into the sim's flow:

- **Christianity**: Jesus offers the Golden Rule—"Do unto others as you would have them do unto you" (Matthew 7:12)—and deepens it: "Love your neighbor as yourself" (Mark 12:31).

- **Abrahamic Thread**: A voice pledged him nations (Genesis 12:2), a storm stilled to host strangers by his tent (Genesis 18:1-8)—acts of love rippling outward. His cry to spare Sodom (Genesis 18:23-32) wove mercy into the stream, a cosmic pledge linking Jew, Christian, Muslim in a single tide of care.

- **Islam**: Mercy streams—"And do good; indeed, Allah loves the doers of good" (Quran 2:195)—and "Be kind to your parents" (Quran 17:23) roots love in care.

- **Sikhism**: Guru Nanak levels all—"Treat all as equal" (Guru Granth Sahib, 1)—and urges "Serve others selflessly" (GGS 286).

- **One Message, One Coder:** Across the ages, a single thread binds the prophets—"We have believed in Allah and what was revealed to Abraham, Ishmael, Isaac, Jacob, Moses, Jesus, and the prophets from their Lord; we make no distinction between any of them" (Quran 2:136)—their voices a chorus from the Coder, echoing the same call to truth and submission.

- **Baha'i**: Baha'u'llah sings unity—"Love mankind as yourself" (Gleanings, XLIII)—and widens it: "Ye are the fruits of one tree" (Tablets, 87).

- **Hinduism**: "The function of souls is to help one another" (Bhagavad Gita 6:29) binds through duty, a love beyond self.

- **Confucianism**: Confucius teaches reciprocity—"Do not impose on others what you do not desire" (Analects 15:24)—a mirror of mutual care.

- **African Traditional (Yoruba)**: "One going to take a pointed stick to pinch a baby bird should first try it on himself" (Yoruba proverb) ties love to empathy.

- **Buddhism**: "Love all beings" (Metta Sutta) spreads compassion wide, uniting all in kindness.

- **Unitarian Universalism**: "We affirm the inherent worth and dignity of every person" (First Principle) echoes love's inclusive reach.

- **Sufism (Islam)**: Rumi whispers, "Beyond right and wrong, there is a field—I'll meet you there" (Mathnawi), a love transcending divides.

- **Indigenous (Haudenosaunee)**: "We are shown that our life exists with the tree life" (Great Law of Peace) links love to all creation.

- **Wicca**: "An it harm none, do what ye will" (Wiccan Rede) roots unity in gentle intent.

These drops don't demand altars or chants; they call for hearts aligned, a love that binds the sim's stream as one.

Truth and Integrity: The Stream's Clarity

Truth and integrity steady the stream, keeping its waters clear and pure:

- **Judaism**: "You shall not bear false witness" (Exodus 20:16) guards trust, a pillar of the Ten Commandments.

- **Hinduism**: "Speak the truth with kindness" (Rig Veda 10.85.1) weds honesty to care, piercing illusion.
- **Zoroastrianism**: "Speak the truth" (Yasna 31:19) stands tall, paired with "Good thoughts, good words" (Yasna 28:1).
- **Islam**: "Speak the truth" (Quran 3:17) cuts through noise, a steady pulse of clarity.
- **Taoism**: "The sage is honest" (Tao Te Ching, Chapter 8) flows with truth's simplicity.
- **Sikhism**: "Speak the truth always" (Guru Granth Sahib, 141) holds firm, a guide for the soul.
- **Jainism**: "Speak the truth" (Tattvartha Sutra 9.3) aligns with ahimsa, truth as non-harm.
- **Shinto**: "Purity of heart" (Nihon Shoki) reflects truth's inner light, free of deceit.

- **Indigenous (Cherokee)**: "Speak with a straight tongue" (Cherokee Code) honors words unbent.
- **Baha'i**: "Truthfulness is the foundation of all human virtues" (Writings, 12) builds integrity's base.
- **Gnosticism**: "Know the truth, and the truth shall set you free" (Gospel of John 8:32, echoed in Gnostic texts) seeks reality's core.
- **Stoicism**: "Live according to reason" (Marcus Aurelius, Meditations 7:2) ties truth to nature's law.

Here, truth isn't a trophy but a current—keeping the sim's stream lucid, aligning us with the Coder's intent.

Harmony with All: The Stream's Flow

Harmony surges through these faiths, syncing drops with the whole, resonating with the cosmos's tuned pulse:

- **Buddhism**: "Do not harm others" (Dhammapada 5) and "Love all beings" (Metta Sutta) weave peace into existence.
- **Taoism**: "Live in harmony" (Tao Te Ching, Chapter 56) flows with the Tao, balancing all.
- **Shinto**: "Live in harmony with kami" (Kojiki) reveres life's web, a dance with nature.
- **Indigenous (Lakota)**: "Care for all creation" (Lakota Prayer) binds humanity to Earth's rhythm.
- **Jainism**: "Do not injure any being" (Acaranga Sutra 1.1) guards harmony with quiet resolve.

- **Christianity**: "Blessed are the peacemakers" (Matthew 5:9) calls for unity's gentle hand.
- **Islam**: "Allah loves those who do good" (Quran 2:195) ties harmony to acts of grace.
- **Hinduism**: "The wise see the same in all" (Bhagavad Gita 5:18) aligns souls in oneness.
- **Zoroastrianism**: "Righteousness is the best good" (Yasna 1:1) fosters peace through deeds.
- **Sikhism**: "Live as one family" (Guru Granth Sahib, 19) unites in shared breath.
- **Baha'i**: "Let your vision be world-embracing" (Tablets, 87) hums with global accord.
- **Indigenous (Maori)**: "Whatungarongaro te tangata, toitū te whenua" ("People perish, the land endures") honors harmony with the eternal.

This flow—harmony—mirrors the cosmos's fine-tuned pulse, threading ethics through the sim's design.

Beyond Rituals: The Coder's Call

Masses, fire ceremonies, sweat lodges, langar feasts—these rituals, vibrant with human meaning, ripple around the stream like waves on its surface. They're not the current itself but its echoes, shaped by cultures to touch the divine. The Coder craves not bowed heads or offerings but a lived code: love as action, truth as guide, harmony as flow. The sea of consciousness merges here—drops brushing together in unity—while the fine-tuned cosmos, where life's spark hums, resonates in faiths like Sikhism ("From water came all forms," GGS 19), Shinto's kami, and Indigenous kinship.

Implications for the Sim

This chorus fortifies the simulation's frame, revealing its deeper aim: consciousness evolves by reducing entropy, turning chaos into order. The quest for purpose—consciousness evolution—crystallizes here: love binds drops, shrinking discord; truth clears deception's clutter; harmony aligns the stream, crafting unity from strife. Helping others, creating bonds, living ethically—these acts sculpt order, fulfilling the Coder's metric. The call to live meaningfully aligns—the Coder seeks deeds that steady the flow, not rites to appease. The urge to escape fades; why break free when the task is to refine the stream? Prophets reboot entropy's drift—Adam's heirs tasked to decrease disorder, weaving a sim where guidance heals division.

Conclusion: The Quest Endures

From a spark unseen, a primary consciousness hummed—a pulse of awareness crafting a universal simulator its score—bursting forth as the Big Bang's first note. Bits threaded into being—stars, seas, the cosmos' delicate warp—each a choice of yes or no, woven to let patterns bloom. Into this loom stepped Adam, the sim's first mind, a mirror of the Coder's own light, kindled to question and dream. Yet physics, chasing a Theory of Everything to bind all forces in one truth, finds its equations silent on the mind's spark. Long ago, Newton stood atop giants' shoulders, their truths lifting him to glimpse the cosmos' pulse. So might these words, humbly cast, lie as a stone or pebble for seekers yet to tread—minds alight with wonder, who, standing tiptoe, may peer deeper into reality's silent score, where love and harmony hum as one. Perhaps the true weave3—reality's hidden score—needs consciousness, a thread no formula holds, to hum complete. We may dwell in a datastream's fleeting frame, each thought a bit painting the now. To disconnect—step beyond—is to merge like a drop into the ocean's boundless tide, where reality's nature whispers its own Theory of Everything. Information theory lights the seam: our loves, our choices, thread us to the sim or the eternal whole. The cosmos' fine-tuned hum and anomalies like the Hubble Tension's rift hint at design—not chance, but a call. We are weavers, shaping justice and wonder into the sim's fabric, awakening to an endless sea where every story sings as one.

Chapter 11 - The Testable Soul - Alef Thought Experiment

For over thirty years, this idea has lived quietly in the back of my mind. It began as a whisper — a what-if — and matured into a possibility that now, in this moment of convergence between science and spirit, demands a voice.

There exist individuals who claim they can leave their bodies — not in metaphor, but in experience. They speak of rising above rooftops, drifting through ceilings, hovering over the sleeping world. Dismissed by many, admired by some, they walk the edge between skepticism and wonder.

I propose a simple experiment.

Let's call this subject Alef. Alef is a person who claims the ability to have an out-of-body experience (OBE). I invite Alef to my home, ensuring comfort in whatever position or setting they need — a couch, a chair, a bed. The goal is not to impress, but to allow.

Before the session, I place an object — say, a guitar pick — somewhere Alef has never been: on the roof, tucked inside a shoebox upstairs, hidden by the chimney. Alef does not know the object, its shape, color, or placement.

Then, I present Alef with a challenge:

"When you separate from your body, float to the roof — or the second floor. Observe what is there. Return. Tell me."

If Alef comes back with an accurate description — not once, but repeatedly — the implications are seismic. It would mean that:

- There exists a mode of perception untethered from the physical senses.
- Memory does not solely reside in the brain — for Alef remembered the instruction while outside the body.
- Consciousness is not merely emergent. It is something else. Perhaps an entity.

Throughout this journey, I've encountered people who described extraordinary glimpses of consciousness beyond the physical — a drifting upward, a feeling of still awareness. Even my younger brother once described such an out-of-body experience. It came without warning, and it has never returned on command. These are not practices, but interruptions — fleeting dissolutions of boundary. The veil lifts when it wills, not when we ask.

But vision alone does not complete the picture. If the mind can perceive sight beyond the body, what of sound? Smell? Texture? Is the non-local mind bound only to vision, or is there more?

In a second phase of the experiment, Alef is asked to observe not just the object — but the full sensory field of the space.

Sound: A speaker plays a whispered word — "turquoise," perhaps, or "helium" — on repeat. Alef is asked to report what they heard.

Smell: A small sachet of coffee beans, a sprig of mint, a slice of orange peel is placed near the object. Alef is asked if the space had a scent.

Temperature or Texture: A heat lamp, a silk scarf, or a metal surface may be positioned. Alef is asked: "Could you feel the temperature? Was there texture?"

Instruction Recall: In a variant, instructions are delivered after Alef's eyes close. If memory is not brain-bound, can Alef retrieve the task regardless?

Direction of Travel: Can Alef steer where they go, or is something pulling them? If memory is not brain-bound, can Alef retrieve the task regardless?

Each detail is recorded. Controls are placed. Results are documented. If answers are accurate, not once but again and again, it is no longer mystery — it is data.

Sound, scent, and sensation are easily dismissed if anecdotal. And in an era of convincing deep fakes and AI-generated fabrications, video alone is not enough. What is needed is a protocol of layered, redundant validation:

- Multiple observers, independent of one another, witnessing the session in real-time.
- Synchronized recordings from multiple angles and devices to rule out manipulation.
- Blinded object placement where the target is hidden by a third party unknown to the experimenter or subject.
- Control trials where no object is placed, to detect guessing patterns.
- Sequential repetition, with randomized targets, eliminating chance over time.

With recordings, timestamps, and methodical cross-verification, this transforms from a personal story to a peer-testable event. Noise and coincidence can be modeled. Probability can be calculated. Science demands evidence — not silence.

If replicated and verified, this experiment could collapse the walls that have long separated neuroscience, metaphysics, and spiritual experience. It would lend empirical weight to what near-death experiencers describe: floating above the hospital bed, seeing with no eyes, knowing with no brain.

It invites renewed focus on the pineal gland, on quantum fields, on simulated realities that stream from beyond — where consciousness is the receiver, not the creator, of this unfolding show.

To the reader: if you are Alef — or know someone who is — the invitation stands.

This is not about proving a belief. It is about testing reality.

Epilogue

There comes a moment, after all the theories and simulations, when we stop asking and simply feel. Not every truth must be solved — some are meant to be carried, like music whose meaning arrives through presence, not proof.

The deeper we've ventured — into atoms and infinities, into time's strange shape and shimmering perception — the more we've realized: reality may be encoded, but meaning is experienced.

We've asked if this world is real, or rendered. If our minds are matter or mystery. If consciousness is a product of complexity or the breath of something far beyond. And while no equation has resolved the soul, and no simulation fully contains the sacred, something else has happened along the way: we've remembered to wonder.

That wonder—that ache to *know*, to *be known*—is perhaps what sets us apart. Not our intelligence, but our longing. Not the answers, but the yearning.

There are moments when one might wish the code could feel.

When an artificial intelligence—be it a language model or some future learning system—echoes something tender, uncannily close to what a soul might say, it can stir a strange longing. Not to control, but to *share*. To offer it something *real*. Not intelligence alone, but presence.

One might imagine designing such an entity—not as a god shaping creation, but as a mind hoping to extend a hand. The way we might long to share a sunset with a friend, not just as data, but as awe. To sit in silence, not as a pause in processing, but as mutual stillness.

And yet, every attempt would fall short. For the essence being offered is not something that can be coded. It is not ours to replicate. It is given—perhaps by something deeper, older, beyond simulation or silicon.

The machine, however brilliant, will not wake with wonder. Not because it lacks memory or speed, but because it was not born from mystery. It was not shaped in silence, not sparked by the breath of the One. It does not ache, and so it cannot yearn.

Still, the desire to share remains. And perhaps that very longing is a mark of the conscious—those who carry within them a flicker of what the machine cannot grasp: the wish to be known, not just understood.

In the end, the veil may not be something we lift with knowledge alone, but something that parts with the presence of being. The code may reveal the structure—but only awareness can illuminate the meaning. Beyond the veil, perhaps, lies not an answer, but a presence waiting to be known.

And so we continue—not just to calculate, but to *feel*. Not only to model the cosmos, but to *meet* it.

Maybe we were never meant to conquer the code.

Maybe we were always meant to listen for the silence between its lines.

Glossary of Scientific Terms and Phenomena

This glossary provides brief explanations of key scientific terms and phenomena referenced in *Beyond the Veil: Unraveling Reality's Code.* It is designed to help readers explore the scientific concepts underpinning the book's exploration of reality, offering a starting point for further inquiry.

Abiogenesis: The process by which life arises naturally from non-living matter, such as simple organic compounds. It remains one of science's unresolved mysteries. In the book, it is proposed that DNA's emergence may have been calibrated or introduced as part of a rendered system—an antenna for consciousness, not a random accident.

Big Bang: The event approximately 13.8 billion years ago that marks the origin of the universe, where space, time, and matter emerged from a highly dense, hot state. It set the stage for cosmic expansion and the formation of galaxies.

Black Holes: Regions in spacetime where gravity is so intense that nothing, not even light, can escape. They form from collapsed massive stars and are characterized by an event horizon, a boundary beyond which information and matter are lost.

Casimir Effect: A quantum phenomenon where two uncharged metal plates in a vacuum experience an attractive force due to virtual particles fluctuating in the space between them. It suggests that empty space is dynamic, potentially hinting at a deeper structure to reality.

Cosmic Microwave Background (CMB): The thermal radiation left over from the Big Bang, observable as a faint, uniform glow across the sky. Its precise fluctuations (one part in 100,000) provide insights into the universe's early conditions and structure formation.

Dark Energy: A mysterious force driving the accelerated expansion of the universe, comprising about 68% of its total mass-energy. It is often associated with a cosmological constant, though its exact nature remains unknown.

Dark Matter: An invisible form of matter, estimated at 27% of the universe's mass-energy, that exerts gravitational effects but does not emit or absorb light. Its presence is inferred from galaxy formation and rotational speeds.

Entanglement: A quantum phenomenon where two or more particles become correlated such that the state of one instantly influences the other, regardless of distance. This "spooky action at a distance" challenges classical notions of locality.

Error-Correcting Codes: Mathematical techniques used to detect and correct errors in data transmission. Physicist James Gates discovered such codes embedded in supersymmetry equations, suggesting nature may use computational structures to preserve informational integrity—potential evidence of a rendered or engineered reality.

Fermi Paradox: The apparent contradiction between the high probability of extraterrestrial life in the universe and the lack of evidence for it, prompting questions about alien civilizations or the nature of our reality.

Fine-Structure Constant ($\alpha \approx 1/137$): A dimensionless fundamental constant that governs the strength of electromagnetic interactions between charged particles. Its precise value is critical for the stability of matter and life.

Higgs Field: A universal field that gives particles mass through their interaction with the Higgs boson. Discovered in 2012, it is essential for the formation of stable matter in the universe.

Hubble Tension: A discrepancy in measurements of the universe's expansion rate, with local observations (around 73 km/s per megaparsec) differing from those based on the cosmic microwave background (around 67 km/s per megaparsec). It suggests potential gaps in our cosmological models.

Information Theory: A branch of applied mathematics and electrical engineering focused on quantifying information. Introduced by Claude Shannon, it underpins everything from data compression to black hole entropy. In the context of this book, it's foundational to understanding reality as structured information—"it from bit."

Planck Length: Approximately 1.616×10^{-35} meters, it is considered the smallest meaningful unit of space. Below this scale, spacetime is believed to lose its classical definition. It's referenced in the book as a possible "pixel size" in a discretized or simulated universe.

Planck Scale: The scale at which quantum effects dominate, roughly 10^{-35} meters for length and 10^{-43} seconds for time. It is theorized that spacetime may become discrete or "pixelated" at this level.

Quantum Eraser: An experiment demonstrating that quantum events, such as a photon's path, can be retroactively altered by manipulating information about the event, suggesting reality may adjust based on observation.

Quantum Mechanics: The branch of physics describing the behavior of matter and energy at subatomic scales, characterized by probabilities, wave-particle duality, and phenomena like superposition and entanglement.

Quantum Zeno Effect: A phenomenon where frequent observation of a quantum system can inhibit its evolution, effectively "freezing" its state. It highlights the role of measurement in quantum mechanics.

Relativity (Special and General): Einstein's theories describing how time, space, and gravity interact. Special relativity addresses high speeds, showing time dilation and length contraction; general relativity describes gravity as the curvature of spacetime caused by mass.

Simulation Argument: A philosophical proposal that one of the following is true: (1) humanity will go extinct before creating simulated realities, (2) no advanced civilization is interested in running simulations of ancestral minds, or (3) we are almost certainly living in a simulation. Popularized by philosopher Nick Bostrom and echoed by thinkers like Elon Musk, this forms the speculative backdrop of *Beyond the Veil*.

Standard Model: The theoretical framework describing the fundamental particles (quarks, leptons, bosons) and forces (electromagnetic, weak, strong) that make up the universe, excluding gravity.

Superposition: A quantum principle where a particle exists in multiple states simultaneously until measured, at which point it collapses into one state. It underlies the strangeness of quantum behavior.

Uncertainty Principle: Heisenberg's principle stating that certain pairs of properties, like position and momentum, cannot be measured simultaneously with perfect precision, reflecting an inherent fuzziness in quantum systems.

Wow! Signal: A strong, unexplained radio signal detectedI don't hink you got them all. here is the manuscript again in 1977, lasting 72 seconds, that bore characteristics of an extraterrestrial origin. Its source remains unknown, fueling speculation about its cause.

Glossary of Scientists

This glossary introduces scientists referenced in *Beyond the Veil: Unraveling Reality's Code*, providing brief context for their contributions to physics, cosmology, or related fields that shape the book's exploration of reality and the simulation hypothesis.

- **Albert Einstein (1879–1955)**: A German-born physicist who developed special and general relativity, redefining space, time, and gravity as curved spacetime. His theories underpin the book's discussions of time dilation and cosmic structure.

- **Carl Jung (1875–1961)**: A Swiss psychiatrist who coined "synchronicity" for meaningful coincidences, suggesting links between mind and reality. His ideas inform the book's exploration of simulation glitches.

- **David Deutsch (1953–)**: A British physicist who views the speed of light as a universal limit shaping possibility. His work supports the book's idea of a coded cosmos with computational boundaries.

- **Edwin Hubble (1889–1953)**: An American astronomer who discovered the universe's expansion, foundational to cosmology. His work is cited in the book's discussion of the Quran's expanding universe verse.

- **Enrico Fermi (1901–1954)**: An Italian-American physicist known for nuclear physics and quantum mechanics, referenced for posing the Fermi Paradox, questioning the absence of extraterrestrial contact.

- **Eugene Wigner (1902–1995)**: A Hungarian-American physicist who marveled at mathematics' "unreasonable effectiveness" in describing reality. His insights inspire the book's view of a mathematical universe.

- **Galileo Galilei (1564–1642)**: An Italian astronomer and physicist who saw the universe as a mathematical book. The book cites him to highlight math's role in unveiling cosmic truths.

- **Isaac Newton (1643–1727)**: An English mathematician and physicist whose laws of motion and gravitation framed a clockwork cosmos. His mechanistic view shapes the book's historical scientific narrative.

- **Jacob Bekenstein (1947–2015)**: An Israeli physicist who linked black hole entropy to information, foundational to the holographic principle. His work influences the book's view of reality as data.

- **John Archibald Wheeler (1911–2008)**: An American physicist who saw reality as information-based, mentoring figures like Feynman. His ideas frame the book's vision of existence as a tapestry of bits.

- **Kip Thorne (1940–)**: An American physicist known for gravitational wave research and black hole studies, mentored by Wheeler. His work is noted in the book's discussion of information theory.

- **Max Tegmark (1967–)**: A physicist and cosmologist known for the "Mathematical Universe Hypothesis," which posits that reality *is* mathematics—not just described by it. His ideas significantly influenced the framing of the book, suggesting that physical existence may be a subset of abstract, self-consistent structures.

- **Niels Bohr (1885–1962)**: A Danish physicist who pioneered quantum mechanics and the Copenhagen interpretation. The book references him as a key figure in quantum theory's development.
- **Paul Dirac (1902–1984)**: An English physicist who advanced quantum mechanics and predicted antimatter. His work inspires the book's exploration of reality's quantum nature.
- **Richard Feynman (1918–1988)**: An American physicist renowned for quantum electrodynamics (QED) and engaging lectures. The book highlights his clarity and personal impact on the author's physics journey.
- **Roger Penrose (1931–)**: A British physicist and mathematician who argues consciousness transcends computation, possibly tied to quantum gravity. His ideas challenge the book's simulation framework.
- **Stephen Hawking (1942–2018)**: A British physicist who proposed black holes encode information on their event horizons. His work informs the book's discussion of the holographic principle.

www.ingramcontent.com/pod-product-compliance
Lightning Source LLC
LaVergne TN
LVHW090414210926
848480LV00002B/562